Power Systems

Electrical power has been the technological foundation of industrial societies for many years. Although the systems designed to provide and apply electrical energy have reached a high degree of maturity, unforeseen problems are constantly encountered, necessitating the design of more efficient and reliable systems based on novel technologies. The book series Power Systems is aimed at providing detailed, accurate and sound technical information about these new developments in electrical power engineering. It includes topics on power generation, storage and transmission as well as electrical machines. The monographs and advanced textbooks in this series address researchers, lecturers, industrial engineers and senior students in electrical engineering.

Power Systems is indexed in Scopus

More information about this series at http://www.springer.com/series/4622

Jingyang Fang

More-Electronics Power Systems: Power Quality and Stability

Springer

Jingyang Fang
Duke Univesity
Kaisetslautern, Germany

ISSN 1612-1287 ISSN 1860-4676 (electronic)
Power Systems
ISBN 978-981-15-8589-0 ISBN 978-981-15-8590-6 (eBook)
https://doi.org/10.1007/978-981-15-8590-6

This Springer imprint is published by the registered company Springer Nature Singapore Pte Ltd.
The registered company address is: 152 Beach Road, #21-01/04 Gateway East, Singapore 189721, Singapore

Preface

This book, focusing on power quality and stability, serves as a textbook as well as a reference book for introductory of more-electronics power systems. The book aims to provide senior undergraduate and graduate students as well as electrical engineers with the knowledge and skills needed in analyzing and controlling modern power systems. To facilitate understanding, each chapter introduces basic concepts and phenomena before going into detailed mathematical descriptions, which are paired with simulation and experimental verifications. On top of that, individual chapters are designed to be self-consistent and comprehensive so that readers can directly refer to their topics of interest. In addition to knowledge transfer, this book points out emerging and promising research directions so that researchers can refer to the references thereof.

The book contains five chapters. Chapter 1 provides the basic knowledge of more-electronics power systems. Chapters 2 and 3 are related to power quality. Specifically, Chap. 2 introduces power quality problems, which are addressed in Chap. 3. Similarly, Chaps. 4 and 5 aim to identify and resolve stability problems, respectively. The following paragraphs briefly introduce individual chapters.

Chapter 1 focuses on fundamental knowledge and background of more-electronics power systems. It begins with a brief introduction to power engineering. Next, the chapter presents an overview of power electronics. Subsequently, the chapter presents a bird's eye view of power systems. Finally, the chapter ends with the introduction of more-electronics power systems.

Chapter 2 reviews power quality problems in more-electronics power systems. The chapter first discusses voltage quality problems and their mechanisms. Next, it presents current quality problems, combined with their features. Finally, this chapter gives typical standards on power quality.

Chapter 3 provides state-of-the-art solutions to power quality problems in more-electronics power systems. It starts with passive solutions. This is followed by the major power electronics-based power conditioning equipment, including Uninterruptable Power Supplies (UPSs), Dynamic Voltage Restorers (DVRs), Static Compensators (STATCOMs), Active Power Filters (APFs), and Unified Power Quality Conditioners (UPQCs).

Chapter 4 summarizes stability problems in more-electronics power systems. First, the chapter explores converter-level stability problems caused by improper controller designs, time-delays, weak grids, and power quality problems. Next, the chapter proceeds to system-level stability problems, which should be blamed for active and reactive power imbalances.

Chapter 5 concentrates on stability improvement techniques in more-electronics power systems. In the converter level, the chapter introduces stability improvement techniques according to proper controller design, time-delay reduction, weak grid techniques, and power quality conditioning. In the system level, the chapter demonstrates promising techniques that balance active power through damping and inertia enhancement. In addition, reactive power balance techniques are covered.

In preparation of this book, the author would like to extend his sincere gratitude to his parents and advisors who give enormous help and support!

Durham, USA Jingyang Fang

Contents

Chapter 1
Fundamentals of More-Electronics Power Systems

1.1 A Brief Introduction to Power Engineering

Before going into details of more-electronics power systems, let us first identify the location of our subject on the tree of knowledge. Starting from the trunk of power engineering in this section, we will gradually move on to the branches of power electronics and power systems in the next two sections. After extensive introductions, we will proceed to explore our major subject—more-electronics power systems.

Ever since junior middle schools, we have been studying physics. Among the major branches of classic physics (including acoustics, optics, electricity, thermal physics, and mechanics), electricity remans as an essential yet hot direction during the past 200 years, particularly in the twentieth century. Maxwell's equations and Hertz's experiments clearly validate the existence of electromagnetic waves, which cover not only electricity but also sound and light. By generalization of Maxwell's equations, Yang-Mills gauge theory unifies three fundamental forces, except for gravitational forces, into electromagnetic forces, and thereby further demonstrating the importance of electricity in physics. Theoretical and practical advancements of electricity in the 20th century have revolutionized our daily lives.

If computer science/engineering itself stays as one discipline, electricity mainly involves two disciplines—power (or electrical) engineering and electronic engineering. Broadly speaking, power engineering treats electricity as energy. In this sense, power engineering focuses on production, transmission, conversion, and utilization of electricity. In contrast, electronic engineering pays more attention to information and integrated circuits. Simply speaking, power (or electronic) engineering deals with strong electricity (or weak electricity) that can shock humanity (or safe to operate with). However, the aforesaid features may not strictly be true, and intersections between power and electronic engineering indeed exist. This explains why universities all over the world organize their electric schools or departments

J. Fang, *More-Electronics Power Systems: Power Quality and Stability*, Power Systems, https://doi.org/10.1007/978-981-15-8590-6_1

in various ways. Typical department names include electrical engineering, electronic engineering, electrical and electronic engineering, electrical and information engineering, and electrical and computer engineering, etc.

Despite with variations, our interested topics generally fall within the discipline of power engineering. As such, courses in power engineering (e.g., power electronics) can be necessary and significantly facilitate the study of more-electronics power systems. However, as this book presents from scratch, beginners will also feel comfortable. In general, power engineering comprises five major secondary disciplines—power systems and automation, electric machines and electric apparatus, high voltage insulation engineering, power electronics and power drives, and theory and new technology of electrical engineering. Among them, power systems and automation as well as power electronics and power drives are more closely related to our subject—more-electronics power systems. Therefore, we will separately discuss these two secondary disciplines later. For visualization, Fig. 1.1 shows a tree of knowledge, where the location of more-electronics power systems is clear. Before moving on to the branches of power electronics and power systems, we look at the features of electricity and power engineering.

Electricity is charming in that we can sometimes sense it, such as in the form lightening, but not in most cases. As a result, measurement equipment, e.g., oscilloscopes, becomes necessary as our "eyes". Voltages and currents are two fundamental electrical quantities that can be measured, on the basis of which we may further derive other useful quantities, such as frequency, active and reactive power, and energy, etc. Measurement of electrical quantities has been developed into another discipline, closely related to power engineering. In this book, we will cover some sensing techniques that lay ground for control of electricity. Notably, as electricity transmits at the speed of light, real-time power balancing between generation and demand is another

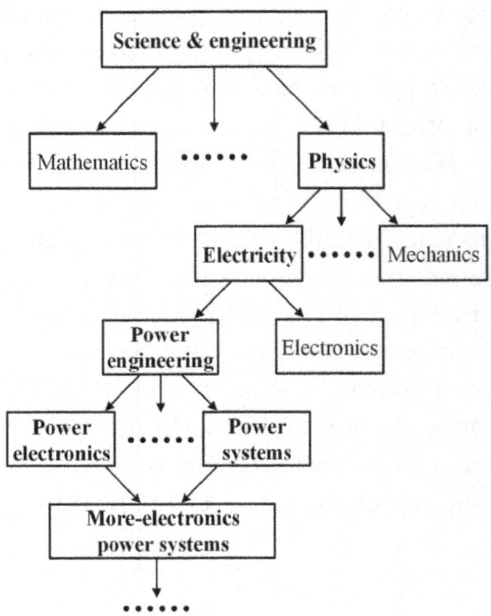

Fig. 1.1 Tree of knowledge

unique feature of electricity. This means that fast control of electricity is essential. In the practice of power engineering, we must keep in mind that electricity can be quite dangerous if not properly handled. Therefore, protection circuits play an important role. Also, heightened attention should be paid to high voltage [e.g., alternative current (ac) voltages >36 V] parts and energy storage elements (such as batteries or capacitors). Last but not least, power engineering highly values the combination of theory and practice, similar to other engineering disciplines. In this sense, concepts and ideas are supposed to be accompanied by experimental validations, which may also be replaced by or paired with simulation verifications for extremely large and/or complicated systems.

1.2 An Overview of Power Electronics

This section focuses on the most active branch of power engineering—power electronics. After introducing the basic concept, we will provide key features of power electronics. In what follows, the section will elaborate on all the subtopics related to power electronics. Finally, we will show typical and emerging applications of power electronics.

1.2.1 What is Power Electronics?

As the name suggests, power electronics covers both power and electronics. Specifically, we control electrical energy through electronic devices. As such, power electronics is essentially an interdisciplinary subject contributed by power, electronics, and control engineering. Dr. William E. Newell, a noted authority of power electronics, first describes power electronics via his famous power electronics triangle, as shown in Fig. 1.2 [1]. Due to its interdisciplinary nature, power electronics involves quite a few of knowledge.

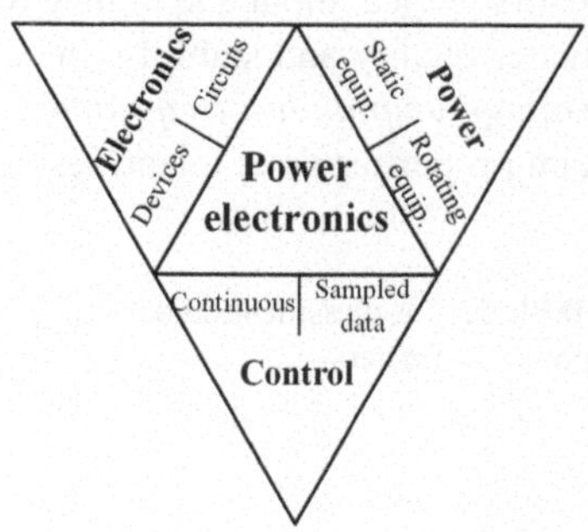

Fig. 1.2 Power electronics triangle invented by Dr. William E. Newell [1]

Fig. 1.3 Vivid example of power converters—a laptop charger

From the practical point of view, power electronics targets at the conversion of electricity from one form to another (e.g., from one voltage to a different voltage). With this regard, we often refer to power electronic equipment as power converters. Figure 1.3 illustrates a vivid example of power converters—a laptop charger. Almost every day, we use laptop and/or cellphone chargers to convert ac voltages from outlets to direct current (dc) voltages for battery charging. Therefore, we may possibly have an intuition of chargers and power electronics. For example, chargers become smaller and lighter in recent years. In addition, they can be hot after working for a while. Such phenomenon reveals some basic yet important features of power electronics, as will be detailed.

As mentioned, laptop and cellphone chargers convert ac voltages into dc voltages. In general, power converters convert either ac or dc voltages to ac or dc voltages. As a result, we have four types of power converters, namely, ac–ac (or ac to ac), ac–dc, dc–ac, dc–dc power converters, as summarized in Table 1.1.

Alternatively, we name ac–dc or dc–ac power conversion as *rectification* or *inversion*, respectively [2]. Conversion between ac and dc electrical quantities is understandable. However, why should we use dc–dc and ac–ac power conversion? Actually, dc quantities feature various amplitudes. For example, voltages of batteries and photovoltaics (PV) panels often differ, thereby necessitating power conversion. Even with identical input and output voltages, power converters may still be necessary for better quality and stability. In ac systems, we need power converters to change *ac voltage amplitudes*, *frequencies*, and/or *phase angles*, which serve as three important properties of ac quantities, as shown in Fig. 1.4. Besides, interconnection of ac

Table 1.1 A classification of power converters

Sources/Loads	Ac ~	Dc –
Ac ~	Ac–ac power conversion	Rectification
Dc –	Inversion	Dc–dc power conversion

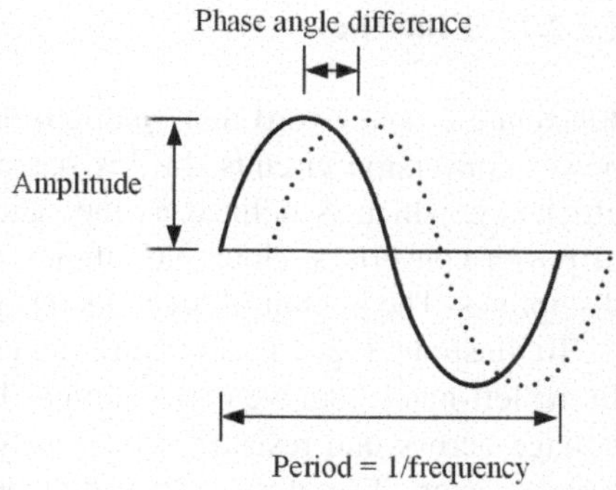

Fig. 1.4 Three important properties of ac quantities

systems with different number of phases (e.g., one-phase and three-phase systems) requires ac–ac power conversion. As a result, power electronics finds its widespread applications, as power conversion is ubiquitous.

It should be kept in mind that not all power conversion is achieved by power electronics, although power conversion is increasingly involving power electronics. One notable example refers to linear power supplies, where semiconductors work as voltage dividers instead of switches. In addition, power electronics excludes the circuits that consist of only passive components, such as resistors, inductors, and capacitors. To better grasp its core, we proceed to key features of power electronics.

1.2.2 Key Features of Power Electronics

In this subsection, we focus on six key features of power electronics. Figure 1.5 presents the keywords related to power electronics. As such keywords will help us to grasp key features of power electronics, we will expand them separately.

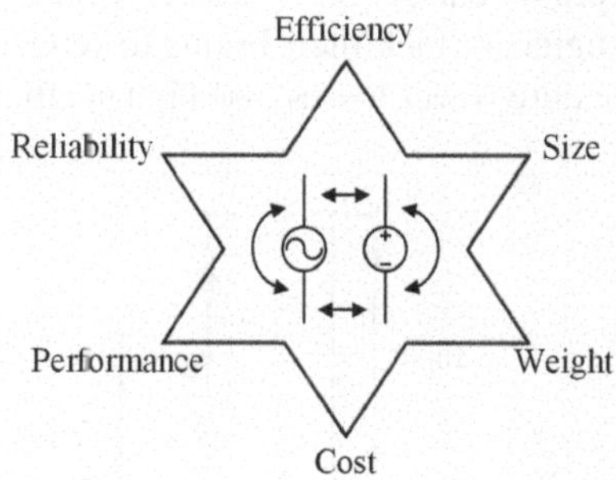

Fig. 1.5 Keywords of power electronics

1.2.2.1 Efficiency

Efficiency is a keyword that distinguishes power electronic converters from other power conversion circuits. In this scenario, efficiency refers to power conversion efficiency, which is defined as the ratio of the output power to the input power of power converters. Note that efficiency may have different implications in other disciplines. The pursuit of *high efficiency* is a key aspect of power electronics.

To illustrate, Fig. 1.6 shows an example of voltage step-down conversion circuits. In the left-hand side, we notice a typical resistive voltage divider, where the output voltage across one resistor is obviously lower than the input voltage across two serial resistors. Therefore, voltage dividers achieve the function of voltage step down. However, resistors cause power losses, particularly in high-power applications, where the currents going through resistors can be large. To get rid of such power losses, power converters generally avoid using resistors in normal operation.

In the right-hand-side of Fig. 1.6, we achieve voltage step down by a buck converter, which consists of two semiconductor switches and an inductor-capacitor (*LC*) filter. First, ignore this *LC* filter and focus on switches. Notably, we do not turn on two switches, simultaneously. Otherwise, the input voltage source will experience undesirable shoot through. Instead, we turn on one switch at a time. With the upper switch on and the lower switch off, the input voltage appears from the output. Alternatively, with the upper switch off and the lower switch on, the output side sees a zero voltage. By changing the ratio of the aforesaid two states in a certain period, we manage to adjust the output voltage. In extreme cases, we obtain an output voltage that equals the input voltage (or zero), when the upper switch remains on (or off). As such, buck converters can only lower voltages. Note that the *LC* filter aims to remove high-frequency ripples of the output voltage.

Ideally, switches cause zero power losses. This is because ideal switches feature zero voltage drops (or currents) when switched on (or off). However, in practice, semiconductor switches do introduce power losses due to nonzero on-state voltage drops, leakage currents, and switching losses, among others. In addition, passive components, e.g., inductors, feature power losses. Power electronic researchers and engineers rack their brains to reduce power losses. On average, power converters feature lower losses and higher efficiency as compared to conventional conversion

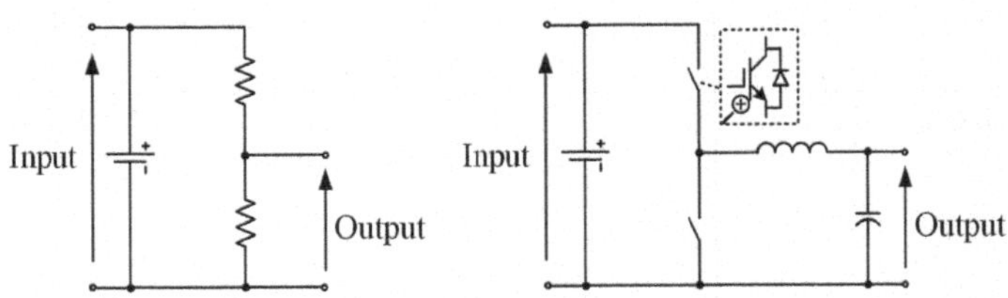

Fig. 1.6 Voltage step-down conversion circuits (left-hand side: a resistive voltage divider, and right-hand side: a buck power converter)

circuits (e.g., resistive voltage dividers). In short, the use of semiconductor switches for high-efficiency power conversion is a key feature of power electronics.

1.2.2.2 Size

Size is another important consideration of power electronics. We target at size reduction. Returning back to the example of laptop chargers, we may have noticed the shrink of their size in recent years. This is a fact of power electronics. In the foreseeable future, power converters will become smaller and smaller.

In terms of size, semiconductor switches and integrated circuits (ICs) are generally smaller than their counterpart—passive components, such as capacitors and inductors. For instance, the *LC* filter in the right-hand-side part of Fig. 1.6 can be large and occupies considerable room in converters. Therefore, research on converter size reduction often focuses on the replacement or removal of passive components.

The strong desire for smaller converter size pushes forward two major directions of power electronics. The first direction lies in the increment of switching frequencies. As mentioned, filters are always necessary to attenuate switching ripples. As the switching frequency increases, switching ripples are easier to remove by smaller filters, thereby leading to the reduction of converter size [3]. However, there is always a trade-off between the increment of switching frequency and efficiency. The reason is that semiconductor switches will introduce intolerable switching losses if they turn on and off too fast. As a result, we have to take care of not switching excessively fast, which limits the improvement of converter size. Fortunately, advances in material science yields new wide-bandgap semiconductor switches, which not only switch significantly faster but also ensure more heat generated by higher power losses. Two notable examples are silicon carbide (SiC) metal–oxide–semiconductor field-effect transistor (MOSFET) and gallium nitride (GaN) FET. Recently, they enjoy growing popularity in the academia and industry. With such modern switches, it is easy to push the switching frequency up to radio frequencies. This explains why we see smaller chargers in the market.

The second direction refers to multiple or multilevel converters. With multiple power converters connected in parallel, we can cancel out ripples, thus leading to the reduction of filter size. In contrast, multilevel converters employ multiple semiconductors switches or power converters in series. Despite with low switching frequencies of individual switches or power converters, the equivalent switching frequency of multilevel converters can increase greatly. For demonstration, Fig. 1.7 presents two common multilevel converters, each consisting of two modules or converter cells (e.g., a buck converter in Fig. 1.6). It should be emphasized that the primary objective of multilevel converters is not on size reduction. Nevertheless, multilevel converters do point out a way of reducing passive components via more semiconductors. On top of passive components, converter size reduction can be achieved through other approaches, e.g., more compact structures.

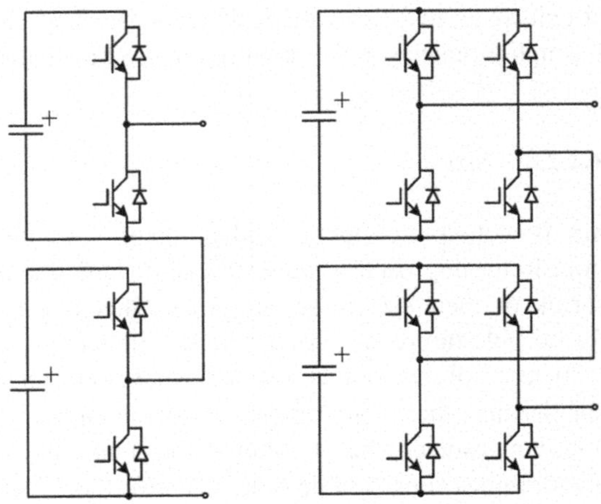

Fig. 1.7 Two common multilevel converters

1.2.2.3 Weight

Weight is a key property of power converters. Despite that system weight and size are two different concepts, they are often closely related to each other. For instance, the increment of switching frequencies often reduces both filter size and weight. Notably, strict requirements on converter weight and size are often prescribed in certain applications, e.g., aerospace. Therefore, we continue to seek for lighter power converters.

With this regard, the terminology *power density* serves as a useful index for evaluation of power converters. Specifically, power density is defined as the ratio of converter power rating and size (or weight), with the unit W/cm^3 or kW/cm^3 (or W/kg or kW/kg). In consequence, the reduction of size and weight translates into an improvement of power density. Similarly, we define *energy density* as the stored energy divided by converter size (or weight).

1.2.2.4 Cost

Cost is important not only in power electronics but also in other engineering disciplines, particularly from the industry perspective. This explains why some perfectly designed power converters cannot survive in the market. In the practice of power electronics, it is often desirable to select components that satisfy objectives instead of those expensive ones with the best performances.

In this sense, we pursue low-cost power converters, which comply with design requirements. In addition to bulky, passive components are often expensive. As such, the removal or simplification of passive components is desirable. This can be achieved by increasing switching frequencies once again. Besides passive components, semiconductor switches, together with their peripheral circuits, may also cost a portion. To this end, we prefer to use fewer switches. Generally, it is better to carefully consider converter cost in the design stage. Otherwise, system retrofitting may significantly increase the overall cost.

1.2.2.5 Performance

The importance of *performances* is obvious. However, this keyword is relatively general under this condition. As specific objectives of individual power converters differ greatly, their key performances can be quite different. Additionally, a single power converter can feature various performances. For example, laptop chargers may require the errors of their output voltages to be within 1%. Meanwhile, their input ac voltages may vary from 100 to 240 V so as to cater to different power grids. These performance requirements must be satisfied. Broadly speaking, requirements on efficiency, size, weight, and cost are also system performances. However, we exclude them from system performances in this scenario, as they are general features of all power converters.

The price paid for the improvement of system performances could be efficiency, size, weight, and/or cost. Dependent on applications, we weigh system performances to different extents. In some cases (e.g., in luxury cars), performances may possibly outweigh costs. We will elaborate performances of power conversion systems throughout this book.

1.2.2.6 Reliability

Reliability receives increasing attention in the power electronic community. Simply speaking, reliability evaluates the ability of power converters to work normally without violating their performance requirements. At the early age, people were indifferent to reliability, as the primary concern lies in normal functions. With large-scale deployment of power converters, reliability becomes hot [4].

Despite of significance, reliability is complicated, as it has tight bearing on all other features of power converters. For example, the extension of converter lifetime may possibly require better components, which in turn enhance system costs. Therefore, trade-offs are often unavoidable. However, smart designs can improve all features of power converters. We always strive for such designs.

In summary, efficiency, size, weight, cost, performance, and reliability are key features of power electronics or power converters. We will be familiar with them in more-electronics power systems, as the book unfolds itself.

1.2.3 Subtopics of Power Electronics

In the previous two subsections, we learn about the basic concept and features of power electronics. This subsection further details the major subtopics of power electronics. Some subtopics are applicable to not only power electronics but also power systems, and hence more-electronics power systems. Therefore, we will elaborate on such subtopics.

Before entering subtopics, let us learn about basic elements of power conversion systems. For illustration, Fig. 1.8 shows a schematic of three-phase inverter systems, which aim to convert a dc voltage into three-phase ac voltages, or vice versa. As compared with Figs. 1.6 and 1.7, Fig. 1.8 provides more details.

First, we notice semiconductor switches and a passive filter (i.e., a three-phase *LCL* filter), together with the dc bus, forming the main circuit of inverters. Second, we find some ideal switches shown in dashed boxes. Some of these switches are connected normally, while others are off. Such switches are symbols of mechanical switches, namely, relays. We employ mechanical switches or relays for protection purposes. Once serious faults occur, relays are tripped off to protect power converters from over currents and/or voltages. In addition, relays enable the change of circuit structures. For example, a negative temperature resistor (NTC) limits dc charging currents during system start up. During normal operation, we turn on the relay S_{dc1} so that the NTC is bypassed for efficiency improvement. Third, we sample electrical signals (e.g., v_{dc1} and i_{cabc1}) through voltage and current sensors (not shown in detail), which are then scaled and filtered, forming the input signals of controllers. Fourth, controllers calculate and output on–off signals of switches and relays, which are amplified by drivers (may also be classified as one part if not included in main circuits) and sent to switches. Note that main circuits, protection circuits, sampling

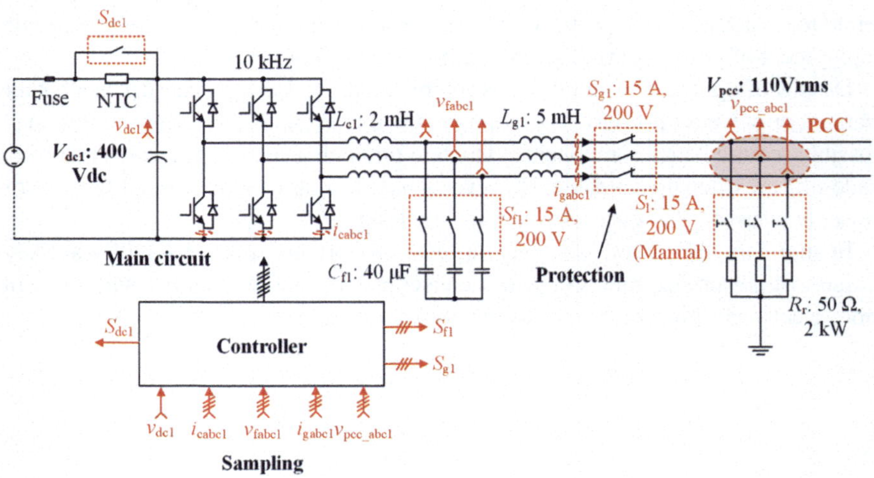

Fig. 1.8 Schematic diagram of inverter systems

circuits, and controllers are essential parts in most power conversion systems. This knowledge will help us study power electronic subtopics.

1.2.3.1 Topology

Topologies refer to circuit diagrams of main circuits. We select converter topologies in the first step of power converter designs. In this sense, a suitable topology can be greatly beneficial. Although there are thousands of different power electronic topologies being invented, most of them have not found practical applications. In other words, the invention of a very useful topology will possibly leave your name in the history of power electronics.

In the early years, dc–dc power converters were the focus of electronic and power electronic researchers. It is not clear who invents the most fundamental topologies of dc–dc power converters, such as the buck converter in Fig. 1.6. However, what we know is that publications on such dc–dc converters appeared before 1970s [5]. To illustrate, Fig. 1.9 presents the topologies of three widely used dc–dc converters, i.e., buck, boost, and buck-boost converters. We further explore their relationships.

By swapping the input and output of buck converters, we obtain boost converters. As a result, boost converters can only increase voltages. Ignoring the input and output voltage sources, we note that the core of buck, boost, and buck-boost converters is a three-terminal cell, consisting of two switches and one inductor. Through rotation of three terminals, we derive boost and buck-boost converters from buck converters. Notably, buck-boost converters may increase or decrease voltages, but they feature reversed outputs [2].

Duality is an important principle for creation of novel topologies. Inductors and capacitors, parallel and serial connection, and voltages and currents are some dual elements. Through duality, we generate Fig. 1.10 from Fig. 1.9. Clearly, input and output voltage sources (or capacitors) are replaced by current sources (or inductors).

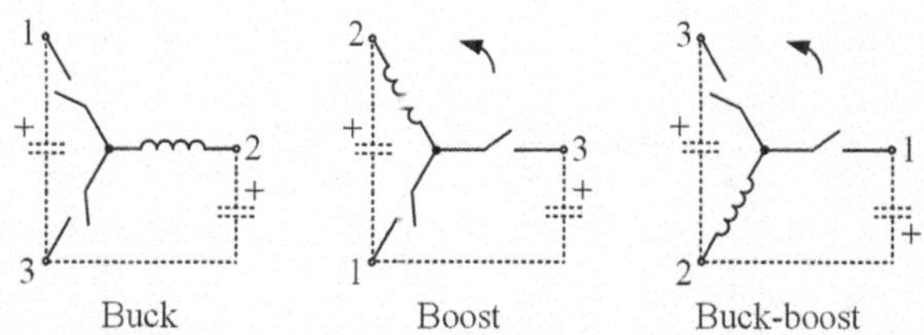

Fig. 1.9 Topologies of widely used dc–dc power converters

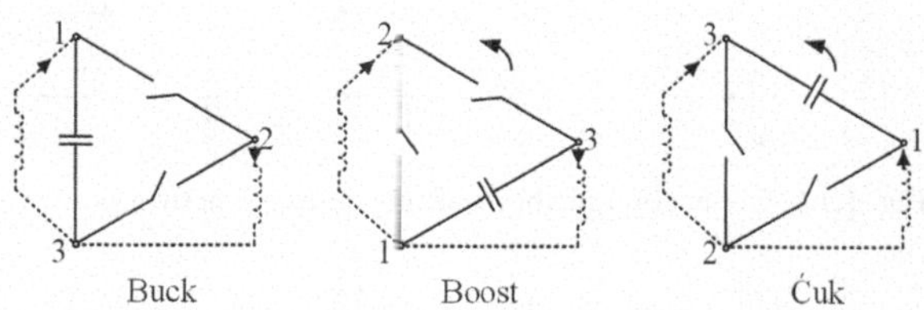

Fig. 1.10 Dual topologies of widely used dc–dc power converters

Moreover, the three-terminal block, consisting of one capacitor and two switches, is connected in a triangular form instead of a Y form. A dual converter of the buck-boost converter was invented and developed by Prof. Ćuk, and hence named the Ćuk converter [6]. Similar to buck-boost converters, Ćuk converters also allow voltage (or currents) step up or down.

Another effective means of creating novel topologies is through inter-connection of multiple converters. As mentioned, we obtain multilevel converters (see Fig. 1.7) by serial connection of converter cells. In general, we represent converter cells as two-terminal or four-terminal networks, as depicted in Fig. 1.11. Remember that three-terminal networks can be regarded as special four-terminal networks.

Further, we can connect networks in series, parallel, or partially serial and partially parallel, yielding novel topologies. For example, Fig. 1.12 illustrates the connection methods of four-terminal networks, including input–output-serial, input–output-parallel, input-parallel and output-series, and input-series and output-parallel methods. With these methods, numerous novel power converters can be derived from the converter cells in Figs. 1.9 and 1.10.

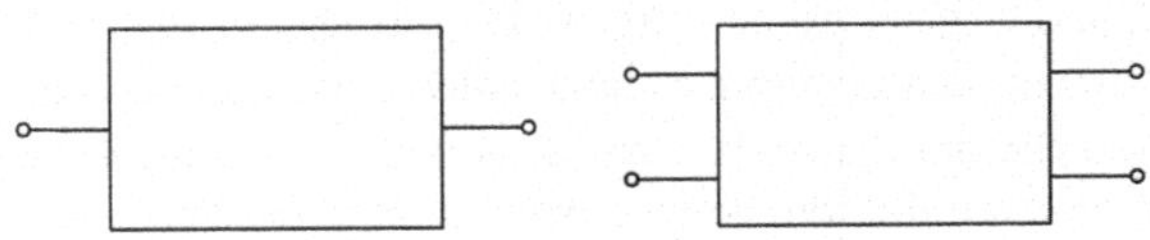

Fig. 1.11 Converter cells as two-terminal or four-terminal networks

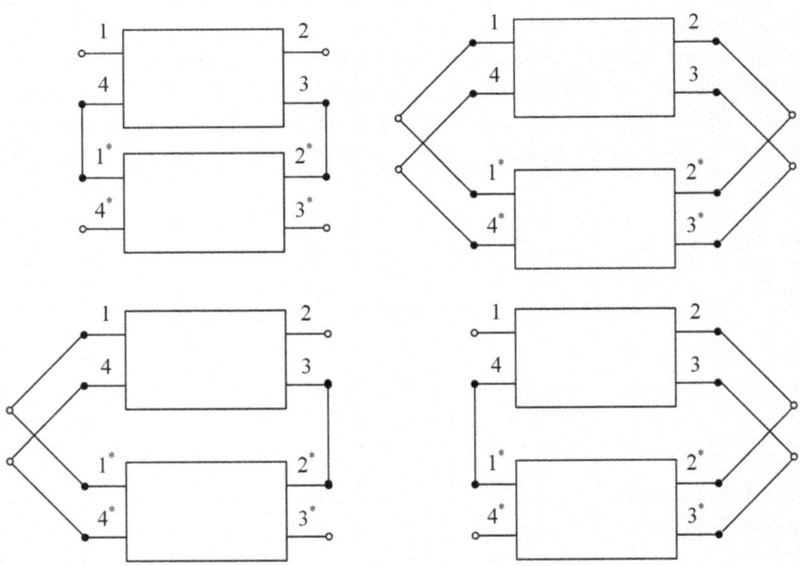

Fig. 1.12 Connection methods of four-terminal networks

1.2.3.2 Modeling

After selecting circuit topologies, we model power converters. *Modeling* lies in the description of power converters by mathematics. Notably, system models are the basis for system analysis and design, which often intertwine with each other. We will detail this part, since modeling is the key for not only power converters but also power systems, and hence more-electronic power systems.

It is worthwhile noting that no model is perfectly accurate. Moreover, accurate models are often at the expense of complexity. Therefore, we should build models according to our needs and avoid overly complicated models. Before modeling, we are supposed to be familiar with three important properties of functions, signals, and/or systems.

Linearity

Linearity should satisfy the superposition principle, which has two sub properties. First, *additivity* requires that a function of the sum of two independent variables equals the sum of their function values, mathematically described by

$$f(x_1 + x_2) = f(x_1) + f(x_2), \tag{1.1}$$

where x_1 and x_2 stand for two random independent elements of any vector space, and $f()$ is a function. Second, the *homogeneity* property states that the function operation and scalar product is exchangeable. That is,

$$f(ax_1) = af(x_1), \tag{1.2}$$

where a is a real number.

In the real world, most systems are nonlinear. However, linearity is a very desirable feature for system analysis and design. Therefore, we tend to model systems as linear systems through linearization of them at their operating points.

Time Invariance

Time variance or *time invariance* describes the relationships of system functions and time. Time-variant system functions change as time goes by. In contrast, time-invariant functions are independent of time. We prefer to time-invariant systems, as they are much easier to analyze and design than their time-variant counterparts. Unfortunately, power conversion systems are all time variant due to their switching behaviors. However, we can make them time invariant by ignoring switching actions and focusing on low-frequency operation, as will be detailed.

Continuity

Continuous systems input and output continuously at all time instants. In contrast, discrete systems feature input and output signals only at a certain time point during a period. Obviously, main circuits of power conversion systems always operate continuously. Therefore, signals of main circuits are continuous. However, as digital

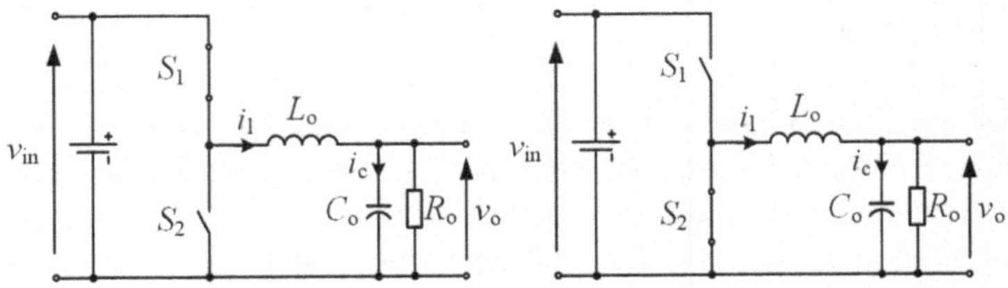

Fig. 1.13 Operating modes of buck converters

controllers receive more attention over analog controllers due to the benefits of higher reliability and robustness, easier update and modification, and better communication capability, we are increasingly dealing with digital and discrete systems. Correspondingly, analog-digital (AD) and digital-analog (DA) signal transformations make a difference on power converters.

Next, we build a simple model of buck converters for modeling demonstration. Figure 1.13 shows the two operating modes of buck converters, where the left-hand-side mode refers to the case of S_1 on and S_2 off, while the right-hand-side mode represents the case of S_1 off and S_2 on. Such segment-continuous behaviors are characterized by power converters.

In the first step of modeling, we use differential equations to describe the relationships among electrical quantities based on Kirchhoff's circuit laws. Generally, quantities of energy storage units, such as inductor currents and capacitor voltages, are necessary and often selected as state variables when performing state-space modeling. Referring to the left-hand side of Fig. 1.13, we have

$$\begin{cases} v_{\text{in}} = L_{\text{o}} \dfrac{\mathrm{d}i_1}{\mathrm{d}t} + v_{\text{o}} \\ i_1 = C_{\text{o}} \dfrac{\mathrm{d}v_{\text{o}}}{\mathrm{d}t} + \dfrac{v_{\text{o}}}{R_{\text{o}}} \end{cases}. \tag{1.3}$$

Similarly, according to the right-hand-side of Fig. 1.13, we derive

$$\begin{cases} 0 = L_{\text{o}} \dfrac{\mathrm{d}i_1}{\mathrm{d}t} + v_{\text{o}} \\ i_1 = C_{\text{o}} \dfrac{\mathrm{d}v_{\text{o}}}{\mathrm{d}t} + \dfrac{v_{\text{o}}}{R_{\text{o}}} \end{cases}. \tag{1.4}$$

By choosing i_1 and v_o as the state variables, v_{in} as the input variable, and v_o as the output variable, we rearrange (1.3) and (1.4) as

$$\begin{cases} \dfrac{\mathrm{d}i_1}{\mathrm{d}t} = -\dfrac{v_{\text{o}}}{L_{\text{o}}} + \dfrac{v_{\text{in}}}{L_{\text{o}}} \\ \dfrac{\mathrm{d}v_{\text{o}}}{\mathrm{d}t} = \dfrac{i_1}{C_{\text{o}}} - \dfrac{v_{\text{o}}}{C_{\text{o}}R_{\text{o}}} \end{cases} \text{or} \begin{array}{l} \dot{\mathbf{x}}_1 = \mathbf{A}_1\mathbf{x}_1 + \mathbf{B}_1\mathbf{u}_1 \\ \mathbf{y}_1 = \mathbf{C}_1\mathbf{x}_1 + \mathbf{D}_1\mathbf{u}_1 \end{array} \text{and} \tag{1.5}$$

$$\begin{cases} \dfrac{di_l}{dt} = -\dfrac{v_o}{L_o} \\ \dfrac{dv_o}{dt} = \dfrac{i_l}{C_o} - \dfrac{v_o}{C_o R_o} \end{cases} \text{ or } \begin{aligned} \dot{\mathbf{x}}_2 &= \mathbf{A}_2\mathbf{x}_2 + \mathbf{B}_2\mathbf{u}_2 \\ \mathbf{y}_2 &= \mathbf{C}_2\mathbf{x}_2 + \mathbf{D}_2\mathbf{u}_2 \end{aligned}, \tag{1.6}$$

where

$$\mathbf{x}_1 = \mathbf{x}_2 = [i_l \ v_o]^{\mathrm{T}}, \mathbf{u}_1 = \mathbf{u}_2 = [v_{\mathrm{in}}], \mathbf{A}_1 = \mathbf{A}_2 = \begin{bmatrix} 0 & -\frac{1}{L_o} \\ \frac{1}{C_o} & \frac{-1}{C_o R_o} \end{bmatrix},$$

$$\mathbf{B}_1 = \begin{bmatrix} \frac{1}{L_o} \\ 0 \end{bmatrix}, \mathbf{B}_2 = \begin{bmatrix} 0 \\ 0 \end{bmatrix}, \tag{1.7}$$

$$\mathbf{y}_1 = \mathbf{y}_2 = [v_o], \mathbf{C}_1 = \mathbf{C}_2 = \begin{bmatrix} 0 & 1 \end{bmatrix}, \mathbf{D}_1 = \mathbf{D}_2 = [0]. \tag{1.8}$$

During a switching period, the two operating modes appear alternately. Assuming that in one switching period T_s, the first mode stays valid from 0 to dT_s, while the second mode holds from dT_s to T_s. Here, d refers to the duty ratio (or duty cycle), which ranges from 0 to 1. In consequence, we can combine (1.5) and (1.6) as

$$\begin{aligned} &\begin{cases} \dot{\mathbf{x}} = \mathbf{A}_1\mathbf{x} + \mathbf{B}_1\mathbf{u} \\ \mathbf{y} = \mathbf{C}_1\mathbf{x} + \mathbf{D}_1\mathbf{u} \end{cases} \quad \mathrm{mod}(t, T_s) \in [0, dT_s) \\ &\begin{cases} \dot{\mathbf{x}} = \mathbf{A}_2\mathbf{x} + \mathbf{B}_2\mathbf{u} \\ \mathbf{y} = \mathbf{C}_2\mathbf{x} + \mathbf{D}_2\mathbf{u} \end{cases} \quad \mathrm{mod}(t, T_s) \in [dT_s, T_s) \end{aligned}, \tag{1.9}$$

where the mod() function performs modulo operation. and the input, state, and output variables are unified. The system model in (1.9) is time variant yet linear in each segment. Another feature is that the operating modes repeat every switching period.

In this stage, we should approach state-space averaging techniques developed by Prof. Middlebrook and Dr. Wester [7, 8]. Their works make huge contributions to power electronics and pioneer converter modeling [2]. The fundamental idea behind state-space average is to average operating states in every switching period. Therefore, in the second step, we average (1.9) on both sides, yielding

$$\begin{cases} \dot{\mathbf{x}} = [d\mathbf{A}_1 + (1-d)\mathbf{A}_2]\mathbf{x} + [d\mathbf{B}_1 + (1-d)\mathbf{B}_2]\mathbf{u} \\ \mathbf{y} = [d\mathbf{C}_1 + (1-d)\mathbf{C}_2]\mathbf{x} + [d\mathbf{D}_1 - (1-d)\mathbf{D}_2]\mathbf{u} \end{cases}, \tag{1.10}$$

which becomes time-invariant but nonlinear, as the duty ratio d is not a constant. Instead, d depends on system states and controllers. It is understandable that (1.10) stays valid only at low frequency bands. With a frequency as high as the switching frequency, either (1.5) or (1.6) is satisfactory, which is different from (1.10). In practice, (1.10) works well below one-tenth of the switching frequency. As such, the first assumption of state-space averaging is *low-frequency analysis and design.*

In the third step, we linearize (1.10) around its steady-state operating point. As a result, the model will be both time-invariant and linear. To do so, we rewrite each variable as a sum of its steady-state and perturbed quantities, yielding

$$\begin{cases} \dfrac{\mathrm{d}(\mathbf{X}+\Delta\mathbf{x})}{\mathrm{d}t} = [(D+\Delta\mathbf{d})\mathbf{A}_1 + (1-D-\Delta\mathbf{d})\mathbf{A}_2](\mathbf{X}+\Delta\mathbf{x}) \\ +[(D+\Delta\mathbf{d})\mathbf{B}_1 + (1-D-\Delta\mathbf{d})\mathbf{B}_2](\mathbf{U}+\Delta\mathbf{u}) \\ \mathbf{Y}+\Delta\mathbf{y} = [(D+\Delta\mathbf{d})\mathbf{C}_1 + (1-D-\Delta\mathbf{d})\mathbf{C}_2](\mathbf{X}+\Delta\mathbf{x}) \\ +[(D+\Delta\mathbf{d})\mathbf{D}_1 + (1-D-\Delta\mathbf{d})\mathbf{D}_2](\mathbf{U}+\Delta\mathbf{u}) \end{cases}, \tag{1.11}$$

where the bolded capital letters represent steady-state vectors, while the variables with the prefix Δ refer to perturbed quantities.

In the fourth step, we ignore perturbations and focus on steady-state relationships, where variable differentials equal zero. In this case, we obtain

$$\begin{cases} [D\mathbf{A}_1 + (1-D)\mathbf{A}_2]\mathbf{X} + [D\mathbf{B}_1 + (1-D)\mathbf{B}_2]\mathbf{U} = 0 \\ \mathbf{Y} = [D\mathbf{C}_1 + (1-D)\mathbf{C}_2]\mathbf{X} + [D\mathbf{D}_1 + (1-D)\mathbf{D}_2]\mathbf{U} \end{cases}, \tag{1.12}$$

Substitution of (1.7) and (1.8) into (1.12), we have

$$\begin{bmatrix} 0 & -\frac{1}{L_o} \\ \frac{1}{C_o} & \frac{-1}{C_o R_o} \end{bmatrix} \begin{bmatrix} I_l \\ V_o \end{bmatrix} + \begin{bmatrix} \frac{D}{L_o} \\ 0 \end{bmatrix} V_{in} = 0 \Rightarrow \begin{cases} V_o = DV_{in} \\ I_l = V_o/R_o \end{cases} \text{and } V_o = V_o, \tag{1.13}$$

where the capital letters refer to the steady-state values of corresponding variables. Clearly, the output voltage V_o is in proportion to the duty ratio D and input voltage V_{in}. As such, we can increase the output voltage by increasing either D or V_{in}.

In the fifth step, we subtract the steady-state relationships described by (1.12) from (1.11), it gives

$$\begin{cases} \dfrac{\mathrm{d}\Delta\mathbf{x}}{\mathrm{d}t} = [(\mathbf{A}_1-\mathbf{A}_2)\mathbf{X} + (\mathbf{B}_1-\mathbf{B}_2)\mathbf{U}]\Delta\mathbf{d} + [D\mathbf{A}_1 + (1-D)\mathbf{A}_2]\Delta\mathbf{x} \\ +[D\mathbf{B}_1 + (1-D)\mathbf{B}_2]\Delta\mathbf{u} \\ \mathbf{y} = [(\mathbf{C}_1-\mathbf{C}_2)\mathbf{X} + (\mathbf{D}_1-\mathbf{D}_2)\mathbf{U}]\Delta\mathbf{d} + [D\mathbf{C}_1 + (1-D)\mathbf{C}_2]\Delta\mathbf{x} \\ +[D\mathbf{D}_1 + (1-D)\mathbf{D}_2]\Delta\mathbf{u} \end{cases}. \tag{1.14}$$

Now, we obtain a linear and time-invariant state-space model, in which all the relevant coefficients can be derived from the steady-state values of state and input variables. In (1.14), the perturbed duty ratio can be lumped together with the input perturbation for further simplification, leading to a standard state-space model. It must be kept in mind that the product of perturbed signals are removed for linearization. Therefore, the second assumption of state-space averaging refers to *small ripples*, covering the ripples of input and state variables as well as the duty ratio.

The state-space modeling methodology introduced in this part applies equally well to other dc–dc converters in Figs. 1.9 and 1.10, and hence excluded here [8]. Heightened attention should be paid to the two assumptions—low frequency operations and small ripples. Recent advances in modeling seek to break these two assumptions. To list a few examples, sampled-data modeling improves model accuracy in high-frequency bands [9]. Large-scale modeling considers the product of perturbed variables [10]. Generalized state-space modeling expands perturbed variables through the Taylor series and then picks the most dominant components [11].

In addition to main circuits, the remaining parts of power conversion systems, such as protection circuits and sampling circuits should also be modeled. Generally, simple models of these parts are sufficient for system analysis and design. However, we have to spend special efforts on their models in demanded applications, such as high-precision power supplies. Under such conditions, AD converters and pulse-width modulator (PWM) drivers may possibly introduce nonlinearity to system models [12]. After modeled, power conversion systems are described by state-space equations or transfer functions, which are ready for further treatment.

1.2.3.3 Control

We will elaborate on this part, as control is extremely useful in both power electronics and power systems, and hence in more-electronics power systems. With system models, power and control experts can readily collaborate on designing controllers. We have two methodologies for analysis and design. The first one refers to the block diagram-based methodology, corresponding to the classic control theory. The second methodology is on the basis of modern control theory and state-space equations, as derived in the previous part.

Classic Control

First of all, let us discuss about classic system analysis and design methodology. In general, power engineers are more familiar with this methodology. Simple and mature are two major benefits. For better illustration, we continue to derive the transfer function from the perturbed duty ratio to the perturbed output voltage of buck converters. First, ignore the input of (1.14) and reorganize it as

$$\begin{cases} \frac{\mathrm{d}\Delta \mathbf{x}}{\mathrm{d}t} = \mathbf{A_d}\Delta \mathbf{x} + \mathbf{B_d}\Delta \mathbf{d} \\ \mathbf{y} = \mathbf{C_d}\Delta \mathbf{x} + \mathbf{D_d}\Delta \mathbf{d} \end{cases}, \text{ in which} \tag{1.15}$$

$$\begin{aligned} \mathbf{A_d} = D\mathbf{A_1} + (1-D)\mathbf{A_2}, \mathbf{B_d} = (\mathbf{A_1} - \mathbf{A_2})\mathbf{X} + (\mathbf{B_1} - \mathbf{B_2})\mathbf{U} \\ \mathbf{C_d} = D\mathbf{C_1} + (1-D)\mathbf{C_2}, \mathbf{D_d} = (\mathbf{C_1} - \mathbf{C_2})\mathbf{X} + (\mathbf{D_1} - \mathbf{D_2})\mathbf{U} \end{aligned}. \tag{1.16}$$

Taking the Laplace transform on both sides of (1.15), we derive

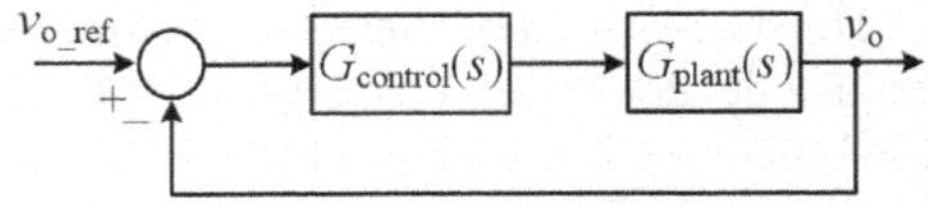

Fig. 1.14 Simplified control block diagram

$$\begin{cases} (s\mathbf{I}-\mathbf{A_d})\Delta\mathbf{x}(s)=\mathbf{B_d}\Delta\mathbf{d}(s) \\ \Delta\mathbf{y}(s)=\mathbf{C_d}\Delta\mathbf{x}(s)+\mathbf{D_d}\Delta\mathbf{d}(s) \end{cases} \Rightarrow \frac{\Delta\mathbf{y}(s)}{\Delta\mathbf{d}(s)}=\mathbf{C_d}(s\mathbf{I}-\mathbf{A_d})^{-1}\mathbf{B_d}+\mathbf{D_d}, \quad (1.17)$$

where s refers to the complex frequency notation. Substitution of (1.7) and (1.8) into (1.16) and then the result into (1.17), we obtain

$$G_{\text{plant}}(s)=\frac{\Delta\mathbf{v}_o(s)}{\Delta\mathbf{d}(s)}=\frac{\frac{1}{C_oL_o}}{s^2+\frac{s}{C_oR_o}+\frac{1}{C_oL_o}}, \quad (1.18)$$

which is the system plant model. With plant models, we can further design controllers. For visualization, Fig. 1.14 shows a simplified control block diagram, in which we ignore all other blocks except for the controller and system plant, modeled as $G_{\text{control}}(s)$ and $G_{\text{plant}}(s)$, respectively. When designing controllers, it is important to know the following performance metrices of dynamic systems.

Stability

First of all, *stability* is a prerequisite of system operation. As reflected by the title of this book, we mainly focus on stability and power quality of more-electronics power systems. Therefore, we deal with stability throughout this book.

In brief, stability refers to the system ability of maintaining normal operation after being disturbed. In this sense, we should differentiate *Lyapunov stability* and *asymptotical stability*. After being disturbed, if system states always remain within a certain boundary, we claim systems to be Lyapunov stable [13]. On the basis of Lyapunov stability, if system states further converge to their equilibrium points, systems are defined to be asymptotical stable [13]. Therefore, asymptotical stable systems are always Lyapunov stable systems, but not necessarily vice versa. For linear systems, Lyapunov and asymptotical stability are equivalent. Their differences are mainly reflected in nonlinear systems due to limit cycles.

For nonlinear systems, there are two Lyapunov methods for stability analysis. We have already introduced the first Lyapunov method, which linearizes systems around their operating points. The second Lyapunov method, also known as the Lyapunov stability criterion or the Direct Method, designs a Lyapunov function $V()$ that is similar to potential energies.

For a nonlinear dynamic system $\mathrm{d}\mathbf{x}/\mathrm{d}t=f(\mathbf{x})$, which satisfies $f(\mathbf{0})=\mathbf{0}$. If $V(x)$ satisfies: (1) $V(x)=0$ if and only if $x=\mathbf{0}$; (2) $V(x)>0$ if and only if $x\neq\mathbf{0}$; (3) $\mathrm{d}V(x)/\mathrm{d}t \leq 0$ (<0 for asymptotical stale) for all $x\neq\mathbf{0}$, we call $V(x)$ a Lyapunov function, and the system is Lyapunov stable. Despite with accuracy, the second Lyapunov method requires a Lyapunov function, which can be difficult to construct in practice. Therefore, we often linearize power conversion systems as mentioned.

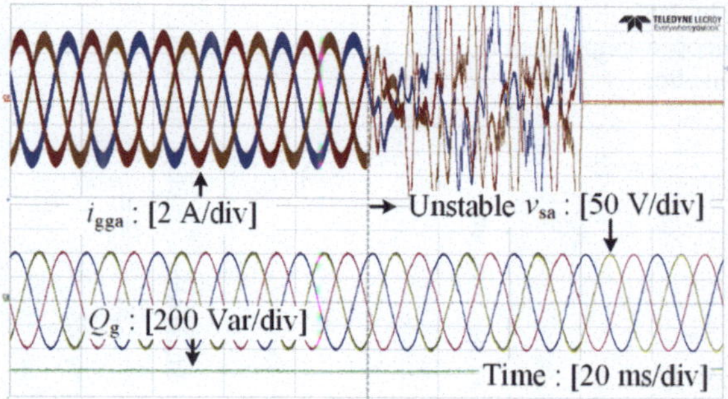

Fig. 1.15 Experimental waveforms of a power conversion system from stable to unstable

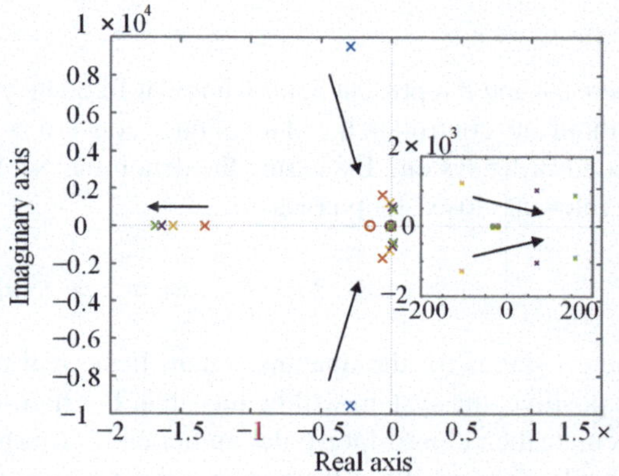

Fig. 1.16 Pole-zero map of a system from stable to unstable

Figure 1.15 presents the experimental waveforms of a power conversion system form stable operation to unstable operation. Clearly, the current waveforms diverge, indicating that the system goes unstable. We should avoid such instability in practice. In essence, all instability problems are caused by positive feedbacks. Such positive feedbacks may occur at certain frequencies, leading to signal amplifications and instability.

According to the classic control theory, we can evaluate stability of linear systems by various approaches, including the Routh–Hurwitz stability criterion, Bode diagrams, Nyquist stability criterion, and root loci, etc. Notably, the most straight forward approach is to calculate all the closed-loop poles. For continuous systems, the poles with positive real parts are unstable poles. For discrete systems, the poles outside the unity cycle cause instability. Figure 1.16 depicts the pole-zero map of a continuous system from stable to unstable, where the right-half-plane poles indicate system instability.

For low-order systems, we can calculate their poles directly. Let us take the standard second-order system as an example, which takes the form of

$$G_{\mathrm{cl}}(s)=\frac{\omega_0^2}{s^2+2\eta\omega_0 s+\omega_0^2}, \tag{1.19}$$

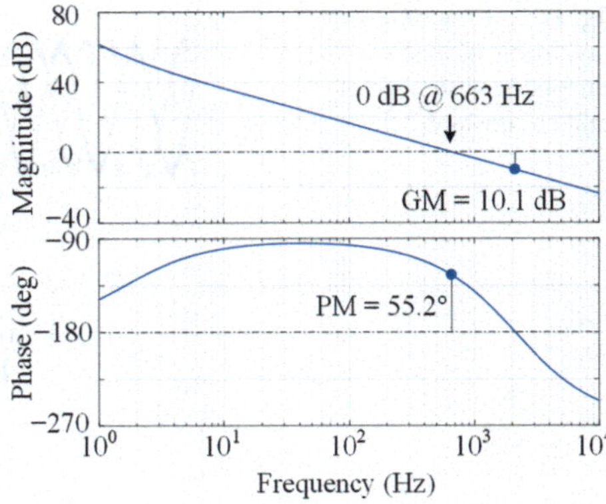

Fig. 1.17 Bode diagram of an open-loop transfer function

where ω_0 and η represent natural angular frequency and damping ratio, respectively. Without any controller, the plant of buck converters in (1.18) is essentially a standard second-order system. By setting the denominator of (1.19) to zero, we can calculate the relevant closed-loop poles as

$$p_{1,2} = -\eta\omega_0 \pm j\omega_0\sqrt{-\eta^2 + 1}, \tag{1.20}$$

where j stands for the imaginary unit. If the real parts of closed-loop poles $-\eta\omega_0$ are positive, the system will be unstable. For high-order systems, it is preferable to calculate their closed-loop poles numerically. This is because analytical solutions are difficult to derive. When the system order is greater than or equal to 5, it is impossible to derive analytical solutions, as the root formula for equations (of 5th and higher orders) does not exist. This is proved by a genius—Mr. Galois [14].

Stability margins are useful for quotative evaluation of system stability, including *gain margins* (GMs) and *phase margins* (PMs), which can be read from open-loop Bode diagrams. For example, Fig. 1.17 shows the Bode diagram of a system. By letting its magnitude equals 0 dB and marking the relevant frequencies, we can obtain the corresponding phase angles. Next, we add the phase angles by 180°, and the results are PMs. Alternatively, by setting the phase equals −180° (in terms of modulus operations), we derive the corresponding magnitudes, whose opposite numbers are GMs. Generally, GMs greater than 3 dB and PMs greater than 30° indicate that system stability can be satisfactory.

Steady-State Errors

Steady-state errors refer to the errors between steady-state outputs and references. We can remove steady-state errors by taking reference signals and controllers into consideration. The core lies in the inclusion of reference signals models in controllers, known as the *Internal Model Principle (IMP)* [15].

Referring back to Fig. 1.14, we assume that the input reference and system loop gain are modeled as

$$v_{o\_ref}(s) = \frac{N_{ref}(s)}{D_{ref}(s)} \text{ and } G_{loop}(s) = G_{control}(s)G_{plant}(s) = \frac{N_{loop}(s)}{D_{loop}(s)}, \tag{1.21}$$

respectively. Notably, we express each transfer function as a fraction, where the numerator and denominator are both polynomial expressions of s.

Through block diagram simplification, we can derive the error signal model as

$$E_{vo}(s) = v_{o\_ref}(s) - v_o(s) = \frac{N_{ref}(s)}{D_{ref}(s)} \cdot \frac{D_{loop}(s)}{D_{loop}(s) + N_{loop}(s)}. \tag{1.22}$$

Furthermore, we derive the steady-state error via the final value theorem as

$$\lim_{t\to\infty} e_{vo}(t) = \lim_{s\to 0} sE_{vo}(s) = \lim_{s\to 0} \frac{sD_{loop}(s)}{D_{ref}(s)} \cdot \frac{N_{ref}(s)}{D_{loop}(s) + N_{loop}(s)}, \tag{1.23}$$

where $e_{vo}(t)$ refers to the time-domain expression of the error signal. Note that IMP is under the assumption of stable systems. That is, all the roots of $[D_{loop}(s) + N_{loop}(s)]$ are with negative real parts. In this case, it is clear that if $D_{loop}(s)$ contains $D_{ref}(s)$ [i.e., $D_{loop}(s) = D_{ref}(s)D_{other}(s)$], (1.23) equals zero, thereby indicating that we can remove the steady-state error. Generally, we cannot expect that the plant numerator $N_{ref}(s)$ will cancel out $D_{ref}(s)$. As such, the controller should include $D_{ref}(s)$ for astatic tracking.

In conclusion, we should incorporate the reference model (specifically, the denominator of reference signals) into controllers. For example, to tightly track a constant reference (which contains $1/s$), we require an integrator (I) or a proportional-integral (PI) controller. Notably, *PI controllers* play a dominant role in practical power converters and power systems.

When facing with sinusoidal references, we use resonant (R) or proportional-resonant (PR) controllers. *R controllers* are modeled as

$$G_R(s) = \frac{\omega_r^2}{s^2 + \omega_r^2} \text{ or } G_R(s) = \frac{\omega_r^2}{s^2 + 2\eta\omega_r s + \omega_r^2}, \tag{1.24}$$

where ω_r represents the resonance angular frequency of R controllers [16]. The damping ratio η aims to adjust the bandwidth of resonant peaks. Comparing (1.19) and (1.24), we find that R controllers are standard second-order systems. *R* controllers are shaped as (1.24) to contain the models of sinusoidal signals, expressed as

$$G_{Sine}(s) = \frac{\omega_r^2}{s^2 + \omega_r^2} \text{ and } G_{Cosine}(s) = \frac{s}{s^2 + \omega_r^2}. \tag{1.25}$$

To track the references consisting of multiple sinusoidal signals, we employ multiple resonant (MR) controllers in parallel. For instance, an *MR controller* that tracks 3rd and 5th harmonics without any error takes the form of

$$G_{\mathrm{MR}}(s) = \frac{3\omega_{\mathrm{o}}^2}{s^2 + 2\eta(3\omega_{\mathrm{o}})s + (3\omega_{\mathrm{o}})^2} + \frac{5\omega_{\mathrm{o}}^2}{s^2 + 2\eta(5\omega_{\mathrm{o}})s + (5\omega_{\mathrm{o}})^2}. \tag{1.26}$$

Finally, we move on to the case of periodical reference signals. In this case, references can be expanded into the sum of a series of sinusoidal signals via the Fourier series as

$$v_{\mathrm{o\_ref}}(t) = \frac{A_0}{2} + \sum_{n=1}^{\infty} A_{\mathrm{n}} \cos(n\omega_0 t - \varphi_{\mathrm{n}}), \tag{1.27}$$

where A_0 and A_n are the Fourier coefficients in the form of amplitudes, and φ_n denote the phase offsets [17]. To achieve accurate tracking, we should employ the controllers that contain the models of all the sinusoidal signals in (1.27). A notable example is the *repetitive controller* developed by a group of Japanese scholars. In their simplest form, repetitive controllers are expressed as

$$G_{\mathrm{Rep}}(s) = \frac{1}{1 - e^{-sT_0}}, \tag{1.28}$$

where $T_0 = 2\pi/\omega_0$ stands for the fundamental period. By setting the denominator to zero, we derive the poles of (1.28) as

$$s = \pm jn\omega_0, \tag{1.29}$$

which comprises the poles of all sinusoidal components. In fact, we can use different controllers as long as the model of periodical references is incorporated. In practice, repetitive controllers are simple and effective options. However, as all the closed-loop poles are on the imaginary axis. Repetitive controllers in their simplest form will cause stability problems. To avoid such problems, we employ stable coefficients and low-pass filters in real repetitive controllers [18].

It is worthwhile to note that the removal of steady-state errors has two implications. The first implication is accurate tracking, as mentioned. The second point refers to disturbance rejection. In this sense, disturbances are regarded as inputs, but the objective of error elimination remains unchanged. We spend a lot of time on this part, as the content will be helpful for further study.

Dynamics

System *dynamics* are important control performances. Typical metrices related to dynamics include overshoot, settling time, and rising time, etc. Figure 1.18 demonstrates a typical step-response plot, where dynamic metrices are highlighted.

Overshoot refers to the maximum amplitude deviation from its steady-state value. Often, overshoot is expressed in terms of percentage. *Settling time* is the time required for systems to achieve their steady-state error bands, in which errors stay within 2% (or 5%). The definition of *rising time* is not strict. As usual, we define *rising time* as the

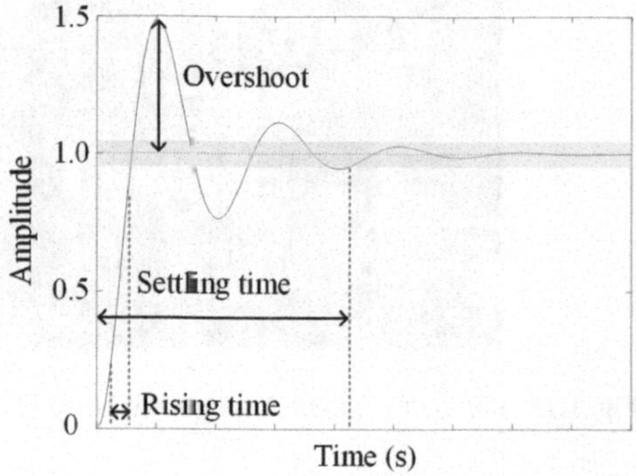

Fig. 1.18 Step-response plot with system dynamic metrices

time duration during which systems reach from 20 to 80% amplitudes. For second-order systems, we have analytical equations for these dynamic metrices. However, for high-order systems, such metrices are often derived via numeric methods, such as simulations. We aim to reduce overshoot, settling time, and rising time. Nonetheless, tradeoffs between dynamics and stability often exist.

Stability, steady-state errors, and system dynamics are three important aspects of controller designs. This holds valid for both classic and modern control theories. As compared to the classic control theory, the modern control theory enjoys the benefits of multi-input multi-output (MIMO) system treatments and manipulation of system states. Next, we briefly discuss the modern control methodology.

Modern Control

According to the modern control theory, systems are built as state-space equations. Correspondingly, analysis and design are performed based on state-space models.

For stability evaluation, we calculate all the eigenvalues of **A**. Systems will be stable if all the eigenvalues of **A** are located in the left-half plane (or within the unity cycle) for continuous systems (or discrete systems) The eigenvalues are of **A** are calculated through the following characteristic equation

$$|\lambda_i \mathbf{I} - \mathbf{A}| = 0, \quad (1.30)$$

where λ_i stand for eigenvalues.

Notably, the techniques developed in the classic control theory can still be applicable to modern control. However, we have multiple inputs and multiple outputs in this case, and hence many transfer functions, mapped between individual inputs and outputs. Therefore, we achieve astatic control only when all relevant transfer functions agree with the IMP.

As long as systems are controllable, we can assign all the eigenvalues as desired through state-feedback control [13]. However, standard state-feedback control requires accurate sampling of all system states, and the relevant sensors and sampling

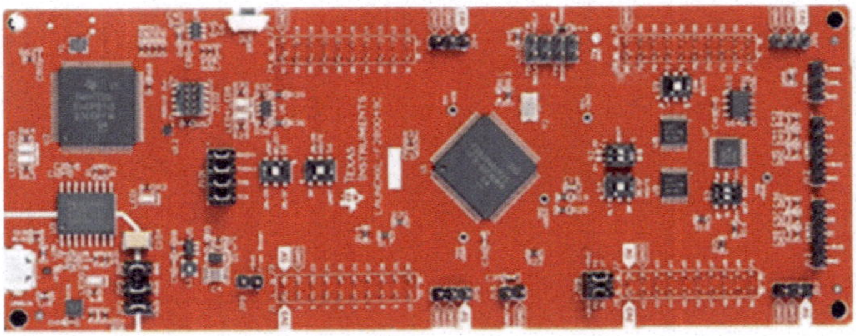

Fig. 1.19 Photo of the DSP development board (TI TMS03XF3F04jD)

circuits add extra cost and complexity to systems. In addition, some states may be difficult or impossible to measure. In that case, we can sample available states and estimate the rest of system states through state observers [13]. Simply put, observers aim to calculate system states through system models.

In addition to the widely used PI and state-feedback controllers, other candidates include on–off controller (or hysteresis controller), robust controller, adaptive controller, one-cycle controller, model predictive controller, fuzzy controller, and neural network controller, etc.

Control Implementations

We have detailed the control theory and performance metrices of control. Next, the practical implementations of controllers will be introduced.

Normally, we use microprocessors to implement control algorithms of power converters. For selection of microprocessors, two popular options are Digital Signal Processors (DSPs) and Field Programmable Gate Arrays (FPGAs). It should be commented that rapid control prototyping (such as dSPACE) becomes popular recently, particularly in the academia, mostly due to its visual interface, hardware simulation, and real-time tuning. However, the cores of rapid control prototyping are normally DSPs and FPGAs. Therefore, we further discuss DSPs and FPGAs.

The operating principles of DSPs and FPGAs differ a lot. Similar to computer CPUs, DSPs execute programs in sequence. As such, the programming language of DSPs (Code Composer Studio) is almost identical to C++. Therefore, DSPs feature strong calculation abilities. In addition, some DSPs are intentionally designed for power electronics and motor drives, where PWM modulators, protection comparators, and sampling filters are included. For illustration, Fig. 1.19 shows a photo of a TI DSP development board (TI TMS03XF3F04jD), where the core DSP is TMS320F28379D [19].

FPGAs consist of numerous logic gates. We program their hardware instead of software. As a result, programming of FPGAs generally takes a longer time than that of DSP. However, due to their hardware implementations, FPGAs allow parallel computing, and hence less latency. Another salient benefit of FPGAs lies in

Table 1.2 A comparison between DSPs and FPGAs

Microprocessors	DSPs	FPGAs
Cost	Low	High
Programming mode	Software	Hardware
Programming speed	Fast	Slow
PWM module	Embedded	NA
Calculation speed	Slow	Fast
IO number	Small	Large
Suitable application	Typical power converters	Multilevel & high-frequency converters

their configurable inputs and outputs, which can easily be altered through hardware programming. To this end, FPGAs are suitable for the applications that require a large number of inputs and outputs (e.g., multilevel converters) and/or very fast control (e.g., high-frequency power converters). Table 1.2 lists a comparison between DSPs and FPGAs.

1.2.3.4 Passive Components

We have elaborated on topology, modeling, and control of power electronics. These subtopics will play important roles in power systems and more-electronics power systems. In the following, we continue to the remaining subtopics. We will not discuss the subtopics in a very detailed manner if they are only related to power electronics itself. This section will introduce main circuit passive components, mainly including inductors and capacitors. Notably, inductors or magnetics remain as an important subarea of power electronics.

Capacitors

Capacitors are necessary in almost all power converters. They aim to filter high-frequency ripples (see the *LC* filter in Fig. 1.6), support voltages as voltage sources (see the dc capacitor in Fig. 1.8), and/or store energy. Electrolytic and film capacitors are two common capacitors used in main circuits.

Generally, electrolytic capacitors feature a higher energy density as well as lower cost and size. As such, they are preferable in voltage support and energy storage applications (e.g., dc voltage support). However, electrolytic capacitors are often polarized, thus excluded from ac circuits. In contrast, film capacitors can undertake voltage stresses from both directions and feature a longer lifetime. However, they are usually bulkier and more expensive. Considering these features, film capacitors often serve as filtering capacitors. Nevertheless, there is a trend of using film capacitors in replacement of electrolytic capacitors for improved converter reliability. To do so, capacitances must significantly be reduced to maintain current converter cost and size [20].

Inductors

Inductors serve as filters and energy storage elements. They are often the bulkiest components in power converters. In addition, the magnetic cores of inductors may cause significant power losses, thus limiting the enhancement of switching frequencies. As such, there is strong desire to reduce the size of inductors or ideally remove them.

Inductors can be made with or without magnetic cores. Without magnetic cores, inductors require much more windings than those with cores. To save copper and efforts, magnetic cores appear in most inductors. As a result, the reduction of size, cost, and weight of magnetic cores becomes one major objective when performing inductor designs. As mentioned, the increment of switching frequencies and the number of power converters are two practical way for such an objective. However, the increment of switching frequencies is limited by switching and magnetic core losses. In contrast, the use of a large number of power converters will increase semiconductors and system complexity.

There are two other methods that can reduce size, cost, and weight of inductors. The first method exploits novel filter topologies [21]. Generally, inductances can be greatly reduced in high-order filters. The prices paid for high-order filters are additional *LC* resonances and complexity. Even though, high-order filters are increasingly popular due to their size, weight, and cost benefits. The second method refers to magnetic integration, where multiple inductor windings share a common magnetic core. Magnetic fluxes generated by different windings cancel out, thus leading to size, weight, and cost reduction of magnetic cores. Figure 1.20 shows an EE magnetic core with coupled inductor windings for size, cost, and weight reduction, where a photo of the magnetic core is also included [3]. However, the mechanism of performance improvements is different from conventional magnetic integration, as detailed in [3]. Power converters, which entirely get rid of inductors, are emerging. Typical examples are switched-capacitor converters.

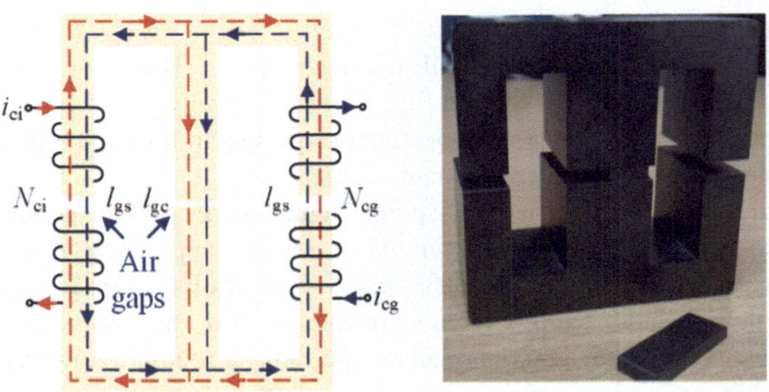

Fig. 1.20 Magnetic core with coupled windings and its photo

When designing inductors and magnetics, finite element analysis software, such as Ansis and Ansoft, can be helpful. Such software allows visualization of electromagnetic field distributions around magnetic cores and/or inductors.

1.2.3.5 Package and Integration

In parallel to passive components, semiconductors switches are of paramount importance in power electronics. Despite that power electronic researchers normally use semiconductor switches designed by material researchers rather than fabricating them, power electronic people design module package and integration.

For sensitive power converters, such as those with high switching frequencies and precision, their layouts must carefully be designed to reduce parasitic parameters (e.g., leakage inductances). This forms the motivation of package and integration. In addition, we may pack switches, together with their protection circuits (or even drivers), into switch modules, resulting in user-friendly switch modules. For instance, half-bridge, single-phase H-bridge, and three-phase modules can easily be found in the market. In addition, the intelligent power modules (IPMs) that integrate protection and driving circuits find growing applications.

For low-power converters, we may integrate not only drivers and protection circuits but also filters, sampling circuits, and controllers into single chips. In this case, power converters become ICs. However, this is not a common practice in high-power converters due to heat dissipation issues. In fact, heat design and mechanical layout are also related to power electronics. However, the relevant discussions are excluded in this book, since they do not belong to electricity.

1.2.3.6 Sampling, Protection, and Drivers

As discussed, sampling, protection, and drivers (if not included in power modules) are important parts of power conversion systems. We will explain these parts. Clearly, sampling circuits aim to transfer signals from main circuits to controllers, while protection circuits and drivers do oppositely in general.

Sampling

Sampling circuits sense information and then scale electrical signals into allowable ranges of controllers. Note that not all information is in the form of electricity. For example, temperature is not an electric property. Therefore, sampling circuits generally involve two parts—*sensors* and *conditioning circuits*. We will discuss sampling with more details due to its generality.

Sensors sense useful information in any form and change it into electrical signals. For instance, an NTC can be regarded as a temperature sensor. When the temperature goes up, the resistance of the NTC drops. Even for voltage and current sensing,

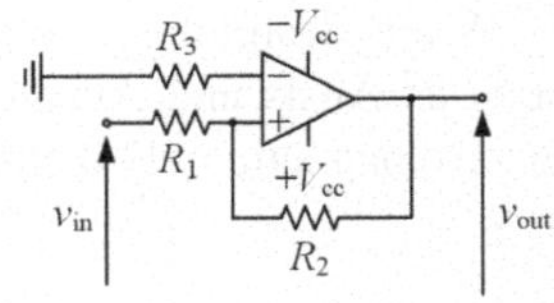

Fig. 1.21 Exemplified signal conditioning circuit

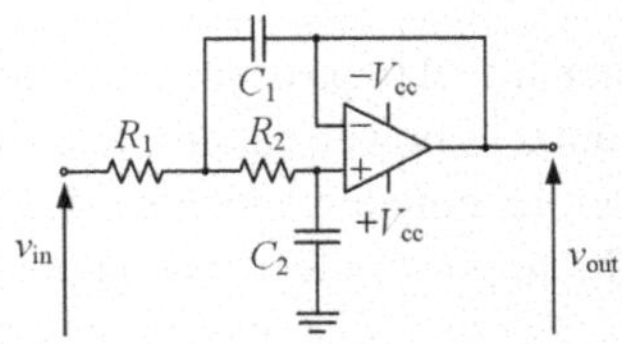

Fig. 1.22 Sallen-Key low-pass filter

sensors are often necessary for isolation purposes. Taking voltage or current halls as examples, we find that they convert high voltages or large currents into low voltage signals. Notice that power converters are not used for voltage and current transformations in this scenario. This is because sensors mainly deal with information instead of energy. Rather than voltage or current sensors, resistive voltage dividers (see Fig. 1.6) are often employed for voltage sensing. In contrast, small sensing resistors are inserted into main circuits for current sensing.

After having electrical signals, we condition them so that they become accessible to digital controllers. Another objective of conditioning is to remove undesirable high-frequency ripples, e.g., switching ripples. Figure 1.21 depicts a signal conditioning circuit, where V_{cc} denotes the reference voltage of operational amplifiers. The resistors R_1 and R_2 achieve the function of signal scaling. R_3 is a balancing resistor.

According to the "virtual short circuit" and "virtual open circuit" features of operational amplifiers, we derive the following input–output relationship:

$$\frac{v_{out}}{v_{in}} = -\frac{R_2}{R_1}. \tag{1.31}$$

Through selection of R_1 and R_2, we can flexibly scale voltages. Notably, we may shift the negative input of operational amplifiers in the case of unipolar controller inputs. Additionally, potentiometers may replace resistors for scaling adjustment.

Besides being scaled (e.g., into 0–3.3 V for DSPs), signals are filtered before entering controllers. In the simplest form, capacitors are used to filter high-frequency ripples. For demanding applications, we use high-order filters. For example, the second-order Sallen-Key filters (introduced by R. P. Sallen and E. L. Key of MIT Lincoln Laboratory in 1955 [22]) are popular choices. They feature high input impedance, low output impedance, high robustness, and simplicity.

Figure 1.22 shows a unity-gain low-pass configuration of Sallen-Key filters. According to the Kirchhoff current law (KCL), "virtual short circuit", and "virtual open circuit", we can derive the relationship between output and input voltages as

$$\frac{v_{\text{out}}(s)}{v_{\text{in}}(s)} = \frac{\omega_{\text{r}}^2}{s^2 + 2\eta\omega_{\text{r}}s + \omega_{\text{r}}^2}, \tag{1.32}$$

which is a standard second-order transfer function. The relevant coefficients are

$$\omega_{\text{r}} = \frac{1}{\sqrt{R_1 R_2 C_1 C_2}} \text{ and} \tag{1.33}$$

$$2\eta\omega_{\text{r}} = \frac{1}{C_1}\left(\frac{1}{R_1} + \frac{1}{R_2}\right). \tag{1.34}$$

The quality factor (Q factor) determines the height and width of resonant peaks, which is defined as

$$Q = \frac{1}{2\eta} = \frac{\sqrt{R_1 R_2 C_1 C_2}}{C_2(R_1 + R_2)}. \tag{1.35}$$

Generally, we design ω_{r} and Q according to applications. Note that there are four variables and two requirements. Therefore, we have two extra design freedoms. By fixing the ratio of resistors and capacitors (such as $C_1 = nC$, $C_2 = C/n$, $R_1 = mR$, $R_2 = R/m$), a one–one relationship is derived as follows

$$\omega_{\text{r}} = \frac{1}{RC} \text{ and} \tag{1.36}$$

$$Q = \frac{mn}{m^2 + 1}. \tag{1.37}$$

In addition to hardware filtering, software filtering becomes popular in recent years. A notable example refers to isolated delta-sigma modulators, paired with digital filters inside controllers [23, 24]. Delta-sigma modulators are compatible with resistor sampling, thus removing expensive sensors (such as voltage and current halls). In addition, they serve as high-precision ADCs with simple conditioning circuits. Voltage and current signals can both be sampled via such modulators.

Protection

As discussed, protection is extremely important for electrical equipment. This holds for both power electronics and power systems. Therefore, we must carefully test protection functions before running power converters and systems.

The fundamental idea of protection is to compare important measured signals with references. If measured signals go beyond certain limitations, we control relays to trip power equipment. Besides, there are also protection circuits (such as fuses and

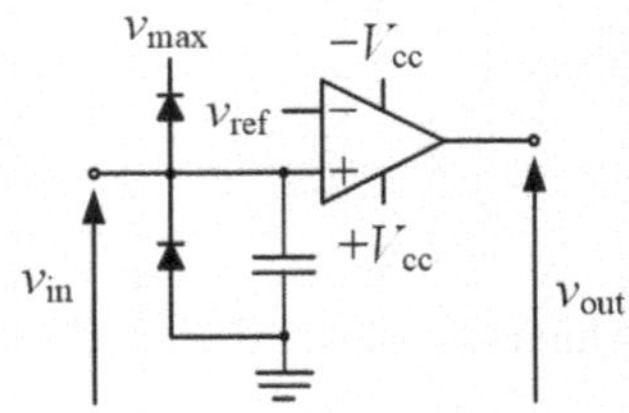

Fig. 1.23 Comparator-based protection scheme

circuit breakers) that trip automatically under faults. In terms of implementations, protection is achieved through either hardware or software.

Figure 1.23 shows a comparator-based protection scheme. As a hardware solution, comparator-based protection is simple and effective. The reference signal v_{ref} can be obtained from either a power supply or a voltage divider. Normally, v_{out} remains negatively saturated. If the input voltage signal v_{in} is greater than v_{ref}, v_{out} will be with a positive value, indicating that faults occur (such as over voltages and currents). Also shown in Fig. 1.23 is a pair of protection diodes, which limit input signals to protect operational amplifiers and their power supplies. It is worthwhile to note that modern controllers may integrate comparator protectors in their chips [24], thus facilitating protection designs.

Software protection has been widely used in replacement of hardware protection to reduce cost and system complexity. However, it should be kept in mind that software protection requires controllers to read signals first. In this sense, software protection is generally slower than hardware protection. Undoubtedly, software protection is more flexible, where digital filtering algorithms can easily be incorporated.

In addition to system-level protection, we also use protection circuits in ICs, particularly for drivers. Many ICs have protection against under- and over-voltage events. On top of that, some drivers have over-temperature, over-current, and shoot-through protection functions. Under faults, they directly lock driving signals so that power converters stop working.

Drivers

Sampling and protection can be found in both power electronics and power systems, but drivers are unique to power electronics. In short, drivers amplify control signals to drive power semiconductor switches.

Drivers are essential interfaces between control and main circuits. They must protect control circuits from strong electricity coupled from main circuits. Alternatively, they have to sufficiently amplify control signals, otherwise switches may not be turned on or off properly. Drivers themselves should undertake high voltages as main circuits. However, they are generally supplied by low voltages and consume low power. Therefore, how to provide high driving voltages form low voltage sources is exquisite.

An interesting way for voltage boost is known as bootstrapping. Figure 1.24 shows

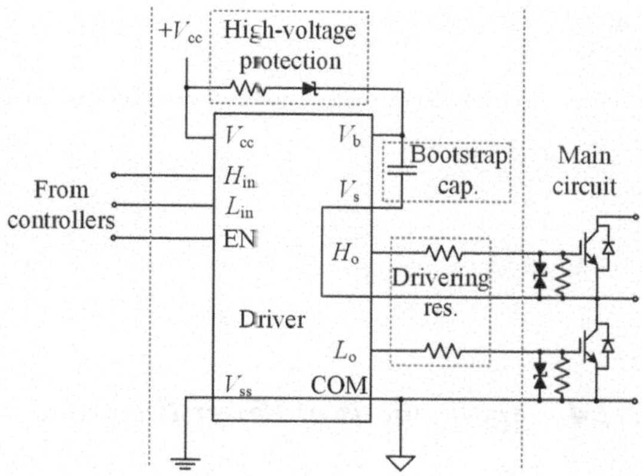

Fig. 1.24 Half-bridge driver with bootstrapping

a half-bridge driver with bootstrapping, where the control signals in the left-hand side are amplified to drive switches in the right-hand side.

First, we look at control signals. H_{in} and L_{in} are PWM digital signals, which are either high or low (indicating on or off). EN stands for the enable signal. Enable signals are common in ICs, which can disable outputs under faults. The driver is supplied by a relatively low voltage V_{cc} (e.g., +15 V). When the lower switch is turned on. Terminals V_s and COM are connected. Through a high-voltage protection diode, the external power supply charges the bootstrap capacitor to V_{cc}. Therefore, bootstrap capacitors only undertake low voltages. Subsequently, the lower switch is turned off, thereby disconnecting V_s and COM. As the lower switch turns off, the potential of V_s increases, and hence the potential of V_b increases, making possible high-side driving. Remember that the low-side switch must be turned on first to charge bootstrap capacitors. Otherwise, drivers cannot work properly. In addition, driving resistors (see Fig. 1.24) should carefully be selected according to switching frequency and electromagnetic interference requirements. This voltage boost technique is similar to the case where a boat goes through a sluice.

Another way is direct driving for the main circuits that are of low-voltage levels. This requires higher voltages supplied by external voltage sources. Generally, external voltage sources should undertake voltages as high as main circuits. In terms of isolation, isolation transformers and optocouplers are two common options. Another notation is that drivers may contain dead-zone generators to avoid shoot through or produce a pair of driving signals from one PWM input signal. In terms of layout, we should ensure gate drive loops as short as possible and avoid placing ground planes near the switching node. Besides, filtering and supporting capacitors are preferably to be very close to drivers. We can refer to driver datasheets for more detailed information [25].

So far, we have overviewed all the subtopics of power electronics. We detailed some of them due to their importance in not only power electronics but also in power systems, and hence more-electronics power systems. Finally, we will cover applications of power electronics.

Fig. 1.25 Photo of electric motors

1.2.4 Applications of Power Electronics

Power electronics is ubiquitous in modern society. It is difficult to summarize all the applications of power electronics. Instead, this subsection presents typical applications of power electronics. In addition, some emerging and promising directions will be pointed out.

1.2.4.1 Classic Applications

Power electronics finds widespread applications in various power supplies. Returning to Fig. 1.3, laptop and cellphone chargers are power supplies. In addition to domestic power supplies, power supplies are everywhere in industry manufacture and production. To list a few examples, telecommunication centers require numerous power supplies. Transportation vehicles and aircrafts require power supplies. Also, induction heating and electrolyte necessitate high-power supplies. Special power supplies find their applications in accelerators and medical equipment. Simply put, power supplies are necessary wherever electricity involves.

Motor drives are one of the main fields of power electronics. The use of power electronics improves the efficiency of motor drives through variable voltage and frequency control. On top of that, power electronics enables bidirectional power transfer and hence saves energy during speed recovery. At the early stage, most power electronic researchers and engineers worked on motor drives. As such, power electronics and power drives collectively serve as one secondary discipline. For visualization, Fig. 1.25 shows a photo of electric motors with clear interior structures.

1.2.4.2 Emerging Applications

We continue to have a view on emerging and promising applications of power electronics. Active research and development are going on in these fields.

Renewable Generation

To reduce carbon footprint and meet the ever-challenging energy demand, renewable energy sources, such as photovoltaics (PV) and wind, are increasingly employed. In most scenarios, renewable energy sources are coupled to power grids through power electronic converters. Therefore, power electronics is essential in renewable generation systems. Alternatively, renewable energy involves a lot of business, and therefore it is undoubtedly one of the most promising direction of power electronics. In fact, the large-scale deployment of renewable energy resources is a salient feature of more-electronics power systems. We will discuss renewable generation in detail later.

Energy Storage

Energy storage is closely related to renewable generation. Its primary objective lies in the compensation of power mismatch between generation and demand in power systems. In addition, energy storage finds applications in transportations, such as vehicles and aircrafts. Common energy storage units include batteries, ultracapacitors, flywheels, fuel cells, and compressed air energy storage, etc. Most of energy storage units necessitate power converters as their interfaces for power and energy management. Additionally, hybrid energy storage systems, e.g., consisting of batteries and ultracapacitors, are also promising [26]. Energy storage is essential in more-electronics power systems, as will be detailed.

Electric and Hybrid Electric Vehicles

Electric and hybrid electric vehicles possess an electric propulsion capability. They produce significantly lower noise and greenhouse gas emissions than conventional fossil fuel-powered vehicles. Apart from these, vehicles powered through electricity feature performance and efficiency improvements. To this end, there is an ongoing trend of electric propulsion in replacement of engine propulsion [27]. Many countries have initiated polices and incentives to speed up the transition. As a frontrunner in the promotion of electric vehicles, Norway registers a 50% market share of electric and plug-in hybrid vehicles now [28]. By 2025, all new vehicles sold in Norway should be zero emission [28]. Power electronics is imperative in electric and hybrid electric vehicles. On top of motor drives, power converters serve as enabling components in vehicle chargers, lighting interfaces, and power supplies for other electric loads. Figure 1.26 illustrates a photo of EV chargers.

Transferring electrical energy to on-board energy storage systems is achieved via either wired or wireless charging. Wireless charging offers a safe, elegant, and convenient solution to the charging of electric without using physical cable connections. Therefore, wireless charging through power electronics emerges as a hot power electronic direction [29].

Fig. 1.26 Photo of EV chargers

1.3 A Bird's-Eye View of Power Systems

In this section, we proceed to power systems. Similar to the introduction of power electronics, we will first learn about the basic concept of power systems. Subsequently, this section will present important features of power systems. Next, the section will introduce the subtopics of power systems. Finally, we will look at typical power systems. Notably, as several related topics, such as control, have been covered in the previous section, they will not further be repeated in this section. However, modeling of power generation will be detailed.

1.3.1 What Are Power Systems?

Generally speaking, a power system is a system consisting of multiple electrical equipment and devices. Conventionally, power systems mainly refer to bulk power grids or utilities, which have been investigated and developed for more than 100 years. However, modern power systems take various forms, such as those in electric vehicles and aircrafts.

In general, bulk power systems are divided into four major parts—generation, transmission, distribution, and utilization [30]. Normally, we buy electricity from power grids every day and use electricity for lightning, heating, and electric appliances, etc. Electricity is generated far away and transmitted or delivered to us through transmission and distribution lines. Therefore, it is easy to find overhead power lines outside. Figure 1.27 shows a photo of overhead power lines. Note that underground cables may replace overhead lines for better reliability.

The four major parts begin to merge in modern power systems, where generators and loads can be exchangeable. For example, electricity customers may also feed power into grid through rooftop PV and energy storage. Figure 1.28 shows a photo of rooftop PV panels. This merge challenges the standard centralized architecture of conventional power systems, which will further be discussed in the next section.

Fig. 1.27 Photo of transmission lines

Fig. 1.28 Photo of rooftop PV panels

1.3.2 Key Features of Power Systems

Next, we concentrate on the key features of power systems. Notably, some features of power electronics are also important features of power systems.

Recapping Fig. 1.5, we note that efficiency, size, weight, cost, performance, and reliability are the key features of power electronics. Indeed, most of them are the key features of power systems. However, we care about these features from different perspectives.

Let us look at efficiency. In power converters, power losses are mostly caused by semiconductor switches and magnetics. In power systems, a considerable amount of power is lost during long-distance transmission. The motivation of efficiency improvement drives the use of high-voltage power transmission. High-voltage power transfer improves system efficiency mainly because of reduced currents under a fixed amount of power, thereby leading to lower power losses and voltage drops across power lines. In terms of efficiency, long-distance power transfer should be avoided if possible. To this end, reactive power compensation is generally achieved locally, as will be detailed. The size and weight of power systems are important aspects. However, the comparison of size and weight of power systems is very difficult. Therefore, we seldom use these indices for power system evaluation. Cost is of great importance in power systems, as transactions of electricity always happen. Therefore, important branches of power systems exist for cost optimization and electricity market design. We will open up this topic later. Performances of power systems vary greatly, but they are important in general. Reliability of power systems is extremely important, as power grid faults may cause huge money or even life losses. In this sense, stable operation is a prerequisite.

In addition to the aforesaid features, power systems are differentiated by the key features as introduced in the following parts.

1.3.2.1 DC, AC, or Hybrid AC/DC

In the history of power systems, there was an intensive conflict between dc and ac systems, also known as the Current War. In the beginning, dc systems were first developed due to their simplicity and convenience. However, people quickly found that dc transmission across a large geographical area became a huge problem. This was mainly due to the lack of dc voltage transformers, and hence high-voltage dc transmission lines. As mentioned, under a certain power rating, low dc voltages translate into high dc currents, leading to significant power losses voltage drops across power lines. In particular, large voltage drops made dc transmission across large areas impossible at that time.

In contrast, people could boost ac voltages through transformers and transmit ac power over long distances. In addition, Mr. Tesla, who is a super genius and pioneer in power engineering, invented practical induction motors and three-phase ac transmission and distribution systems [31]. Induction motors are so successful and robust that they are still in popular and widespread use today, e.g. in electric vehicles. Three-phase systems can transfer more power as compared to single-phase systems when using the same amount of material. Once again, three-phase circuits are dominant in today's bulk power systems. Considering the aforementioned benefits, ac systems eventually beat against dc systems. As a result, the majority of large power systems are ac systems.

However, ac systems are very complicated. As mentioned, ac electrical quantities are characterized by amplitude, frequency, and phase angle. The interaction of ac quantities leads to synchronization problems. For three-phase systems, imbalance is another problem. In contrast, dc electrical quantities only feature their amplitude, and hence simpler. Power electronics enables the flexible change of dc voltages via "dc transformers". As a result, high-voltage dc transmission systems have been developed. Besides, pure dc power systems are promising. However, dc currents feature no zero-crossing points. As a result, faults in dc systems are much more difficult to handle than in ac systems. With this regard, dc circuit breakers are hot topics under active research. Table 1.3 summarizes a comparison between ac and dc power systems.

Even though dc power systems appear to be simple and effective in the long term, the retrofitting of ac power systems will be very difficult and time consuming. As an intermediate process, hybrid ac/dc power grids have been put forward, where dc or ac generators and loads are tied to dc or ac sub grids, respectively [32]. However, it should be kept in mind that ac power systems still play a dominant role, and they will not easily be replaced in the near future. Therefore, this book mainly focuses on power

Table 1.3 A comparison between ac and dc power systems

Power grids	AC	DC
Amplitude	Yes	Yes
Frequency	Yes	NA
Phase angle	Yes	NA
Active power	Yes	Yes
Reactive power	Yes	NA
Voltage transform	Yes	Yes
Protection	Simple	Difficult

quality and stability of ac more-electronics power systems. We will also discuss dc power systems wherever necessary. From another perspective, dc quantities may be regarded as ac quantities whose frequencies are zero.

1.3.2.2 Power Balance

Real-time *power balance* is a salient feature of power systems. As mentioned, electricity transmits at the speed of light. Therefore, we must always ensure that the generated power and demanded power are equivalent.

Power balance includes two aspects—*active power balance* and *reactive power balance*. Active power refers to the power consumed by resistors, while inductors and capacitors only absorb reactive power. In power systems, active power balance is mainly related to frequency control [30]. In contrast, reactive power balance helps maintain grid voltage amplitudes as desired [30]. Power imbalances may cause frequency and/or voltages to go beyond certain limits, thus leading to stability problems. Therefore, system-level stability improvement is essentially fast power balance. We will detail this point later.

Generally, active power can be transmitted over long distances, which is eventually used and paid by electric customers. In contrast, reactive power transmission increases power losses without any payment. As such, power system operators attempt to avoid reactive power transfer and compensate reactive power locally. However, some customers (e.g., motor loads) consume reactive power and introduce additional power losses. As a result, power systems set strict requirements for them in terms of *power factor*—the ratio of active power and reactive power. This necessitates local reactive power compensation, as will be detailed. In addition, energy storage is an important methodology to mitigate the active power mismatch between generation and demand.

1.3.2.3 Resilience

As power systems are the most complicated machines created by humankind, serious breakdowns of power systems may cause huge economic losses. In addition, since bulk power systems are spread over large geographical areas, they are subject to various disasters, such as hurricanes, tsunamis, and/or wars. Therefore, power systems must be prepared for those small-probability yet serious faults. The relevant ability is called power system *resilience*.

Despite its importance, resilience improvement methods are difficult to test. Obviously, experimental validations of them cost a fortune. Instead, simulation validations may not be very reliable due to the limited number of factors being considered. However, resilience methods can be developed cater to individual power systems, whose topologies, models, and control approaches are well known. The use of redundant equipment is a simple way of resilience improvement at the expense of increased system costs.

1.3.2.4 Artificial Intelligence

Another important feature of modern power systems is the widespread use of *artificial intelligence* (AI) and *big data*. AI is suitable for very complex and highly nonlinear systems, such as power systems. Correspondingly, the data involved for training is enormous, thereby requiring big data techniques.

AI and big data are successful in information estimation. For example, equipment monitoring, renewable generation and/or load forecasting, and price estimation, can be achieved by AI. However, despite with intensive research, real-time control through AI and big data has not been proven so far due to slow dynamics. The major problem related to AI is its insecure theoretical basis. Although AI enjoys a great success, the selection of neural network layers, functions, and coefficients are empirical. As a result, we have good results but without knowing the reason behind. The continuous research and development on AI and big data will make them completer and more perfect.

To enable big data, ubiquitous measurement and communication are necessary. A notable example refers to the *phasor measurement unit* (PMU), which measures three important factors of ac quantities, including frequency, magnitude, and phase angle. Through communications, PMUs greatly facilitate power systems in terms of stability analysis and improvement [33]. An obvious trend of modern power systems is the widespread adoption of measurement equipment. On top of measurement, 5G communications add on further benefits.

1.3.3 Subtopics of Power Systems

Power systems involve numerous subtopics, which cannot be elaborated in a single book. As mentioned, several subtopics of power electronics, including topology, modeling, control, sampling, and protection, are also subtopics of power systems. Therefore, we will exclude the repeated contents and focus on important topics related to modern power systems in this subsection.

1.3.3.1 Power System Planning

Power system planning refers to the planning of adding new electrical elements (e.g., generators, loads, and transformers) into existing power systems. The prerequisite of planning is reliable operation of power systems. Therefore, the influences of new elements on the performances of power systems must be investigated. Also, we work out the installation procedure through planning. For example, how to ensure the continuous power supply after grid faults through backup routes should be considered in the planning stage. In this sense, power flow analyses are necessary. Moreover, power system planning is closely related to stability analysis. In addition, forecasting techniques (such as load forecasting) are important in planning.

In general, power system planning can be divided into short-term, medium-term, and long-term planning. Short-term planning deals with contingencies. Medium-term planning focus on equipment retrofitting. Long-term planning updates and reconfigures networks.

1.3.3.2 Optimization

Optimization plays a decisive role in power systems. In fact, optimization exists in every subtopic of power systems. Simply speaking, optimization seeks to choose a best option from multiple options through mathematic calculation.

In an optimization problem, we first have several objectives, e.g., a minimized system cost. Next, we build a function or functions, e.g., $f()$, that quantitatively represent objectives. In consequence, our objectives translate into maximization or minimization of such function(s). Subsequently, we consider some practical limitations, e.g., power outputs of generators. Finally, we solve the optimization problem through mathematical algorithm(s).

To illustrate, let us consider a very simple example of maximum power transfer, as shown in Fig. 1.29, where the input voltage V_{in} and resistor R_i remain fixed. In the first

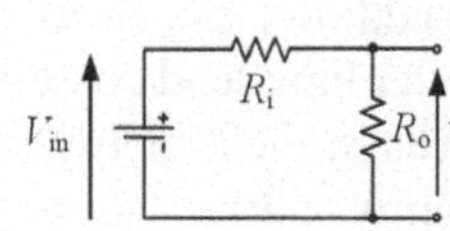

Fig. 1.29 Circuit for optimization of power transfer

step, we select the objective as to make the output resistor R_o consume the maximum active power. Next, we formulate the expression of active power consumption as

$$f(R_o) = \frac{V_{in}^2 R_o}{(R_{in} + R_o)^2}. \tag{1.38}$$

Further, we must ensure that resistances are positive, which forms a limitation. The overall optimization problem is mathematically described as

$$\min_{R_o > 0}\{f(R_o)\}, \tag{1.39}$$

where min() function represents the function of minimization. Finally, using the derivative of (1.38), we derive the optimized R_o and power consumption as

$$R_o = R_{in} \text{ and} \tag{1.40}$$

$$f_{min}(R_o) = \frac{V_{in}^2}{4R_{in}}. \tag{1.41}$$

Generally, we encounter problems that are much more complex than this example. However, the thread of thought remains unchanged. In power systems, we often minimize system costs, among other factors, under the constraints of active and reactive power as well as stability limitations.

1.3.3.3 Energy Market and Policy

Utilities sale electricity through electricity markets. Recently, electricity markets begin to emerge with other related markets, such as heat markets, and combined as *energy markets. Government policies* greatly influence energy markets, and hence should be included.

To avoid monopoly, energy markets across the world have been or tend to be liberalized. Liberalized energy markets include the Australian Energy Market Commission in Australia, the Energy Community in Europe, and the Energy Market Authority in Singapore, etc. [34, 35]. However, as energy is closely related to the safety of countries, totally free and fair energy trade is still impossible in the current world. Even worse, energy problems may lead to wars.

In many countries, multistep electricity price policy has been adopted. It mainly aims to save energy and help power balance. For example, utilities will charge their customers with a higher electricity price if they use electricity beyond a certain limit. In addition, the electricity price under peak load conditions can be higher than that under light load conditions. In addition, most countries promote renewable energies through their energy policies. In this way, companies participate in the renewable energy market can be patronized by the government. Also, electricity generated

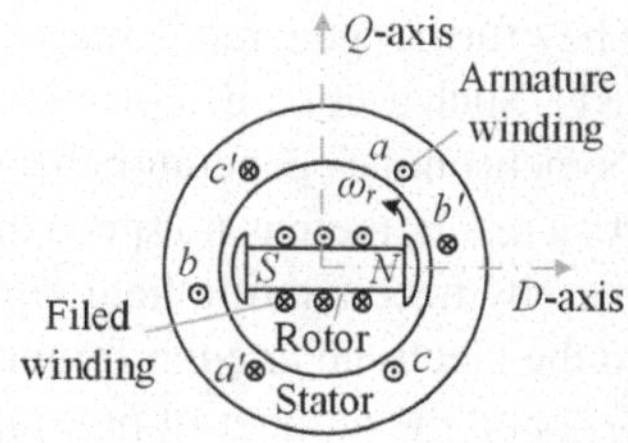

Fig. 1.30 Cross-section view of synchronous generators

through renewable energy sources becomes cheaper. As a result, the relevant energy market promotes. In this sense, energy markets and government policies go hand in hand and cannot be separately considered.

1.3.3.4 Modeling of Power Generation

Power generation is considered as an independent subtopic of power systems due to its significance. This is because the most important control objectives of power systems—active power balance (or frequency control) and reactive power balance (or voltage amplitude control) are both mainly achieved by power generation control. In this part, we will focus on the models of power generation.

Power generation (or electricity generation) refers to the process of generating electricity from primary energy sources. Coal and oil are two conventional primary energy sources. Besides, natural gas is another popular alternative, e.g., in Singaporean power grids [35]. In modern power systems, water, wind, and solar are three important primary energy sources, among others. Generally, power plants generate electricity in two steps. In the first step, we burn coal, oil, or gas in boilers, yielding high-pressure steam. This step is mainly related to thermal engineering, and hence excluded here. In the second step, steam drives turbines, which in turn offers the driving force of synchronous generators (SGs). Next, we focus on the model of power generation, particularly for the models of SGs.

Models of Synchronous Generators

We have to learn about the models of SGs, as they play a very important role on the stability of power systems. However, detailed models of SGs are very complicated, as detailed in [30]. For simplicity, we model the most important parts of SGs.

Figure 1.30 shows a cross-section view of SGs. As shown, SGs consist of stators and rotors, as those of motors shown in Fig. 1.25 Normally, the stators of SGs comprise armature windings, which are connected to main circuits and generate the major magnetic field. On top of armature windings, SGs also contain field windings and often amortisseurs in their rotors, which are excluded from induction motors. In induction motors, the rotational magnetic field generated by armature windings induce rotor voltages, which give rise to rotor currents through closed conductors.

They further generate a magnetic pole that always lags the rotational magnetic field [31]. Such a smart design allows the removal of field windings. Notably, we call SGs "synchronous" generators because they run in synchronism with the grid frequency. As a result, there is no speed difference between the rotor and magnetic field of SGs, thereby necessitating field windings to produce magnetic poles. Note that ω_r refers to the electrical angular frequency of SGs. It is determined by the mechanical speed ω_{sm} and the number of poles p_f as follows [30]:

$$\omega_r = 2\pi f_r = \frac{p_f}{2}\omega_{sm} = \frac{p_f n_{sm} \pi}{60}, \tag{1.42}$$

where n_{sm} represents the mechanical rotor speed with the unit of r/min. Let us take SGs with a pair of (or two) poles as an example, a rotor speed n_{sm} of 3000 r/min yields a frequency f_r of 50 Hz. Similarly, we operate SGs at 3600 r/min to generate 60 Hz electricity.

To model SGs, we should recognize them as coupled mechanical and electrical systems. First of all, let us focus on their mechanical parts with slow dynamics.

Mechanical Models of SGs

Rotors of SGs can be modeled as rigid bodies. In particular, the kinetic energy stored in rotors of SGs takes the form of

$$E_{SG} = \frac{J_{SG}\omega_{sm\_base}^2}{2}, \tag{1.43}$$

where J_{SG} stands for the combined moment of inertia of SGs, and ω_{sm_base} denotes the rated mechanical angular frequency.

When modeling power systems, we treat many electrical quantities in their per unit forms to facilitate calculation and comparisons between systems. To this end, the per unit rotor kinetic energy is defined as a very important concept called *inertia* (inertia constant or coefficient), which can be expressed as [30]

$$H_{SG} = \frac{J_{SG}\omega_{sm\_base}^2}{2VA_{base}}, \tag{1.44}$$

where VA_{base} represents the rated power of SGs. H_{SG} has the unit of second (or s). For typical SGs, it ranges from 2 to 10 s. We will gradually understand the important role of H_{SG} on stability of more-electronics power systems.

According to the second Newton's law of motion in its rotational form, we can mathematically describe the motion of SG rotors as

$$T_m - T_e = \frac{J_{SG}\mathrm{d}\omega_{sm}}{\mathrm{d}t}, \tag{1.45}$$

where T_m and T_e stand for the mechanical torque and electrical torque of SGs, respectively. Notably, torques play the same role in rotational motions as forces play

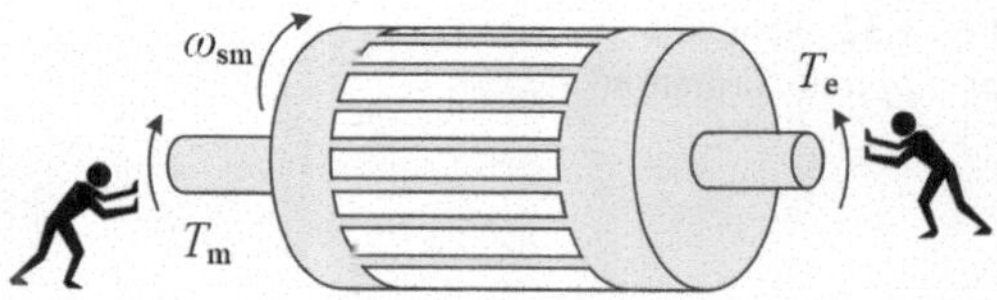

Fig. 1.31 Simplified diagram of synchronous generator motions

in translational motions. For visualization, Fig. 1.31 shows a vivid diagram of SG rotors. The driving mechanical torque T_m is provided by turbines as mentioned. In addition, the electrical torque T_e is determined by electrical loads. The difference between T_m and T_e determines the motions of SGs. Another observation from (1.45) is that the combined moment of inertia J_{SG} tends to slow down SG motions.

Next, we divide (1.45) by the rated power of SGs on both sides, yielding

$$\frac{T_m - T_e}{VA_{base}} = \frac{J_{SG} d\omega_{sm}}{VA_{base} dt} \Rightarrow \frac{T_m - T_e}{T_{ref}\omega_{sm\_base}} = \frac{J_{SG} d\omega_{sm}}{VA_{base} dt} \Rightarrow T_{m\_pu} - T_{e\_pu} = 2H_{SG}\frac{d\omega_{r\_pu}}{dt}, \tag{1.46}$$

where $VA_{base} = T_{base}\omega_{sm\_base}$ is used in the derivation, T_{base} stands for the rated torque, and the subscript pu represents the per unit notation. The per unit motion equation of SGs in (1.46) is the famous *swing equation*. Considering the damping effect offered by SGs' damping windings, the swing equation can be reorganized as [30]

$$T_{m\_pu} - T_{e\_pu} - K_d \Delta\omega_{r\_pu} = 2H_{SG}\frac{d\omega_{r\_pu}}{dt}, \tag{1.47}$$

where K_d represents the damping coefficient, and Δ refers to the perturbation notation. Alternatively, we can reorganize the swing equation as an angle form:

$$T_{m\_pu} - T_{e\_pu} - K_d \Delta\omega_{r\_pu} = \frac{2H_{SG}}{\omega_0} \cdot \frac{d^2\delta_r}{dt^2}, \tag{1.48}$$

where the phase angle δ_r is normally expressed in its real form, which introduces the angular frequency reference ω_0 in the denominator. Sometimes, T_M is employed in replacement of $2H_{SG}$ to represent the mechanical starting time [30].

Synchronous Frame and Rotational Phasor

Before studying SG electrical models, let us first familiar ourselves with the *dq*0-*frame* and *rotational phasor*. In essence, we transform signals from the natural *abc*-frame to the synchronous *dq*0-frame through a nonlinear transformation that has two benefits—the change of ac signals into dc signals and removal of one redundant signal for balanced three-phase systems. Although three-phase ac signals correspond to three-dimensional space phasors, we discuss plane phasors for simplicity.

We define a complex phasor as

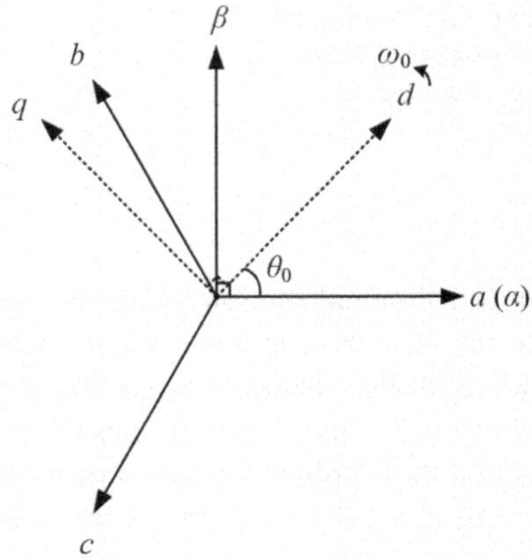

Fig. 1.32 Standard *abc*/*dq* transformation

$$\vec{\mathbf{u}} = U_{\text{ref}} e^{j\theta_0} = U_{\text{ref}} \angle \theta_0, \tag{1.49}$$

where U_{ref} denotes the amplitude of the complex phasor, and θ_0 refers to its phase angle. Note that a phasor can be represented either in an exponential form or an angle form, and the two forms are equivalent [36].

Next, we consider a balanced three-phase circuit, supplying (or loading) armature windings (see Fig. 1.32). Heightened attention should be paid to the location of armature windings, where the phase *b* lags the phase *a* from the rotational direction. Correspondingly, the balanced three-phase voltages are represented as

$$\begin{cases} u_{\text{a}} = U_{\text{ref}} \cos \omega_0 t \\ u_{\text{b}} = U_{\text{ref}} \cos(\omega_0 t - \frac{2}{3}\pi) \\ u_{\text{c}} = U_{\text{ref}} \cos(\omega_0 t + \frac{2}{3}\pi) \end{cases}. \tag{1.50}$$

The first linear transformation, known as the $abc/\alpha\beta 0$ transformation, aims to decouple three axes. The resultant frame is named as the stationary $\alpha\beta 0$ frame. In particular, the α-axis and a-axis are identical, and the β-axis lags the α-axis by an angle of 90° from the rotational direction. It should be noted that the β-axis may also be framed to lead the α-axis by 90°. Correspondingly, the transformation matrices can differ greatly. Without loss of generality, we follow the definitions in Fig. 1.32. Through projection, we derive the voltage components on the respective axes from Fig. 1.32 as

$$\begin{bmatrix} u_{\alpha} \\ u_{\beta} \\ u_0 \end{bmatrix} = \frac{2}{3} \begin{bmatrix} 1 & -\frac{1}{2} & -\frac{1}{2} \\ 0 & \frac{\sqrt{3}}{2} & \frac{-\sqrt{3}}{2} \\ \frac{1}{2} & \frac{1}{2} & \frac{1}{2} \end{bmatrix} \begin{bmatrix} u_{\text{a}} \\ u_{\text{b}} \\ u_{\text{c}} \end{bmatrix}, \tag{1.51}$$

where all the matrix elements can be derived from the phase relationships among the three axes. Taking the α-axis as an example, we observe that the a-axis and α-axis overlap. Therefore, the projection of u_a on the α-axis is itself. Moreover, the c- and b-axes have a $\pm 120°$ phase shift with the α-axis. Therefore, their gains are $\cos(\pm 120°) = -1/2$. Similarly, the other elements can be derived. Notably, we have a gain 2/3, which aims to maintain the amplitudes of signals unchanged. An alternative option is the square root of 2/3, which maintains fixed power during transformation. For balanced three-phase systems, the sum of three-phase voltages equals zero. As such, the 0-axis signal u_0 can be removed. It should be kept in mind that this does not stay valid in imbalanced systems.

The second matrix transformation aims to map sinusoidal signals into dc components. The transformation is nonlinear, as it involves trigonometric operation. Referring to the dq-frame in Fig. 1.32, we notice that it features a phase difference of θ_0 with respect to the stationary $\alpha\beta$-frame. Further, we derive the dq-frame components through the projection of the $\alpha\beta$-frame signals as

$$\begin{bmatrix} u_d \\ u_q \end{bmatrix} = \begin{bmatrix} \cos\theta_0 & \sin\theta_0 \\ -\sin\theta_0 & \cos\theta_0 \end{bmatrix} \begin{bmatrix} u_\alpha \\ u_\beta \end{bmatrix}, \tag{1.52}$$

where the matrix is known as the $\alpha\beta/dq$ transformation matrix given that $\theta_0 = \omega_0 t$. In other words, the matrix in (1.52) is not only nonlinear but also time variant. In addition, note that the 0-axis signal u_0 remains unchanged if incorporated.

Combining (1.51) and (1.52), we derive the standard $abc/dq0$ transformation matrix as

$$\begin{bmatrix} u_d \\ u_q \\ u_0 \end{bmatrix} = \frac{2}{3} \begin{bmatrix} \cos\theta_0 & \cos(\theta_0 - \frac{2\pi}{3}) & \cos(\theta_0 + \frac{2\pi}{3}) \\ -\sin\theta_0 & -\sin(\theta_0 - \frac{2\pi}{3}) & -\sin(\theta_0 + \frac{2\pi}{3}) \\ \frac{1}{2} & \frac{1}{2} & \frac{1}{2} \end{bmatrix} \begin{bmatrix} u_a \\ u_b \\ u_c \end{bmatrix} = \mathbf{T_{abc/dq0}} \begin{bmatrix} u_a \\ u_b \\ u_c \end{bmatrix}. \tag{1.53}$$

Substituting (1.50) into (1.53) and considering that $\theta_0 = \omega_0 t$, we derive that $u_d = u_{ref}$, $u_q = 0$, and $u_0 = 0$. That is, we use a phasor that is aligned with the d-axis to represent the three-phase voltages. The magnitude of the phasor is identical to those of three-phase voltages, while its phase angle is θ_0.

We should be very careful about the dq-frame location in Fig. 1.32, as it varies in different dq-frames. For illustration, Fig. 1.33 shows all possible dq-frames, where abc-frames are removed for clarity. Obviously, we have discussed Fig. 1.33a, which transforms three-phase cosine signals into constants in the d-axis. We continue to look at the remaining cases.

In Fig. 1.33b, the relative location relationship of d- and q-axes remains fixed. However, the q-axis replaces the d-axis in Fig. 1.33a. The corresponding $abc/dq0$ transformation matrix is derived as

Fig. 1.33 Various *dq* frames

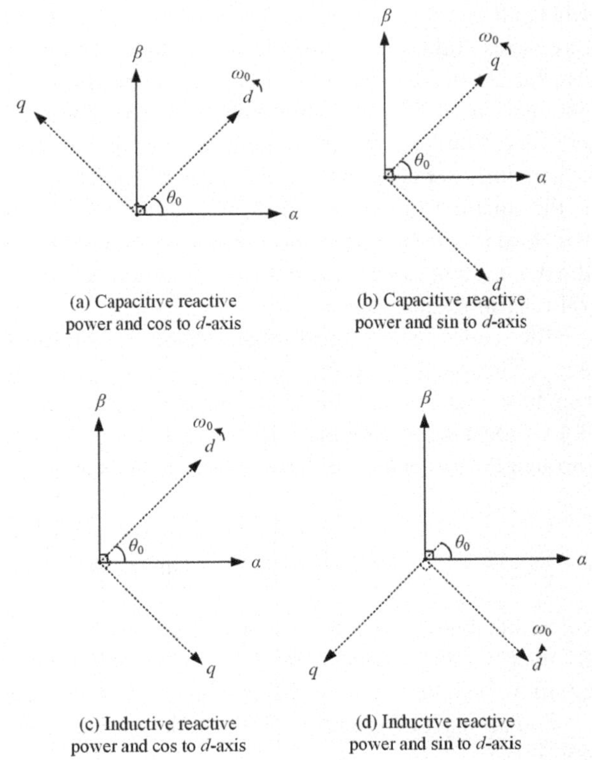

$$\begin{bmatrix} u_{\mathrm{d}} \\ u_{\mathrm{q}} \\ u_0 \end{bmatrix} = \frac{2}{3}\begin{bmatrix} \sin\theta_0 & \sin(\theta_0 - \frac{2\pi}{3}) & \sin(\theta_0 + \frac{2\pi}{3}) \\ \cos\theta_0 & \cos(\theta_0 - \frac{2\pi}{3}) & \cos(\theta_0 + \frac{2\pi}{3}) \\ \frac{1}{2} & \frac{1}{2} & \frac{1}{2} \end{bmatrix}\begin{bmatrix} u_{\mathrm{a}} \\ u_{\mathrm{b}} \\ u_{\mathrm{c}} \end{bmatrix}, \tag{1.54}$$

which transforms a group three-phase sinusoidal signals into a constant in the *d*-axis.

Referring to Fig. 1.33c, where the locations of *d*- and *q*-axes are swapped as compared to Fig. 1.33b. In this case, the *d*-axis leads the *q*-axis by 90°. We derive the related transformation matrix as

$$\begin{bmatrix} u_{\mathrm{d}} \\ u_{\mathrm{q}} \\ u_0 \end{bmatrix} = \frac{2}{3}\begin{bmatrix} \cos\theta_0 & \cos(\theta_0 - \frac{2\pi}{3}) & \cos(\theta_0 + \frac{2\pi}{3}) \\ \sin\theta_0 & \sin(\theta_0 - \frac{2\pi}{3}) & \sin(\theta_0 + \frac{2\pi}{3}) \\ \frac{1}{2} & \frac{1}{2} & \frac{1}{2} \end{bmatrix}\begin{bmatrix} u_{\mathrm{a}} \\ u_{\mathrm{b}} \\ u_{\mathrm{c}} \end{bmatrix}, \tag{1.55}$$

which transforms three-phase cosine signals into constants in the *d*-axis.

Finally, Fig. 1.33d shows the case where sine signals are mapped into constants in the *d*-axis. Accordingly, we derive the transformation matrix as

$$\begin{bmatrix} u_d \\ u_q \\ u_0 \end{bmatrix} = \frac{2}{3} \begin{bmatrix} \sin\theta_0 & \sin(\theta_0 - \frac{2\pi}{3}) & \sin(\theta_0 + \frac{2\pi}{3}) \\ -\cos\theta_0 & -\cos(\theta_0 - \frac{2\pi}{3}) & -\cos(\theta_0 + \frac{2\pi}{3}) \\ \frac{1}{2} & \frac{1}{2} & \frac{1}{2} \end{bmatrix} \begin{bmatrix} u_a \\ u_b \\ u_c \end{bmatrix}. \tag{1.56}$$

Note that the reactive power will be capacitive if the q-axis leads the d-axis (see Fig. 1.33a, b), e.g., i_q leads u_d. Alternatively, the reactive power will be inductive (see Fig. 1.33c, d). In simulations (e.g., under Matlab/Simulink), we can select different *abc*/*dq* transformations. Remember to check such selections when debugging three-phase circuits. Also, make sure the relevant inverse transformations [derived by the inverse matrices of (1.53) to (1.56)] are correctly paired.

Magnetic Quantities Related to SGs

We review some basic magnetic quantities related to SGs. Specifically, the *magnetic motive force* (MMF) generated by windings is defined as

$$MMF_m = Ni, \tag{1.57}$$

where N stands for the number of winding turns, and i represents the winding current. Similar to the electro motive force (EMF), the MMF corresponds to a voltage source in magnetic circuits.

The *magnetic reluctance*, corresponding to the resistor, takes the expression of

$$R_m = \frac{l_m}{A_m \mu_0}, \tag{1.58}$$

where l_m denotes the length of air gaps, A_m represents the cross-section area of magnetics, and μ_0 denotes the permeability of air, which is a constant $4\pi \times 10^{-7}$ N/A^2. Notably, reluctances of magnetic cores are much smaller as compared with those of air gaps. Therefore, R_m normally refers to air-gap reluctances, which depend on shapes of magnetic cores. The reciprocal of reluctances is defined as the *permeance*, i.e.,

$$P_m = \frac{1}{R_m}, \tag{1.59}$$

which is similar to the conductance in electric circuits.

The product of the MMF and permeance is known as the *flux*:

$$\Phi_m = MMF_m P_m = \frac{MMF_m}{R_m}. \tag{1.60}$$

Clearly, fluxes play a similar role in magnetic circuits as currents do in electric circuits.

A unique yet important concept in magnetics is the *flux linkage*, which is defined as the product of the number of winding turns and flux:

Table 1.4 Electrical and magnetic quantities

Quantities	Electrical	Magnetic
Sources	EMF (V)	MMF (A)
Loads	Resistance (Ω)	Reluctance (H^{-1})
Reciprocal loads	Conductance (S)	Permeance (H)
Sources/Loads	Current (A)	Flux (Wb)
Flux linkage	NA	Flux linkage (Wb)
Passive components	Capacitance (C)	Inductance (H)

$$\Psi_{\mathrm{m}} = \Phi_{\mathrm{m}} N. \tag{1.61}$$

Finally, the *inductance* is calculated as

$$L_{\mathrm{m}} = \frac{\Psi_{\mathrm{m}}}{i} = N^2 P_{\mathrm{m}}. \tag{1.62}$$

Table 1.4 summarizes and compares electrical and magnetic quantities, where the corresponding units are included in the brackets.

Real Electrical Models of SGs

Detailed derivations of SG models in terms of electrical and magnetic relationships are complicated. In this book, we mainly follow Prof. Kundur's book (see [30]) and provide the key results in this part.

Following the derivations in the synchronous frame and rotational phasor part, we obtain the synthetic MMF as [30]

$$MMF_{\mathrm{total}} = \begin{bmatrix} \cos\theta & \cos(\theta - \frac{2}{3}\pi) & \cos(\theta + \frac{2}{3}\pi) \end{bmatrix} \begin{bmatrix} MMF_{\mathrm{a}} \\ MMF_{\mathrm{b}} \\ MMF_{\mathrm{c}} \end{bmatrix} = K\cos(\theta - \omega_{\mathrm{r}} t), \tag{1.63}$$

where K refers to the MMF amplitude. Clearly, the resultant MMF rotates counterclockwise.

We ignore the detailed mdeling process of SGs, which can be found in [30]. Through *abc* to *dq*0 transformations, we describe the electrical models of SGs by the following equations:

(1) Stator voltage equations [30]:

$$\begin{cases} e_{\mathrm{d}} = p\Psi_{\mathrm{d}} - \Psi_{\mathrm{q}}\omega_{\mathrm{r}} - R_{\mathrm{a}} i_{\mathrm{d}} \\ e_{\mathrm{q}} = p\Psi_{\mathrm{q}} + \Psi_{\mathrm{d}}\omega_{\mathrm{r}} - R_{\mathrm{a}} i_{\mathrm{q}}, \\ e_0 = p\Psi_0 - R_{\mathrm{a}} i_0 \end{cases} \tag{1.64}$$

where e_{dq0} stand for the stator voltages, Ψ_{dq0} represent the stator flux linkages, i_{dq0} designate the stator currents, p signifies the differential operator, and R_a refers to the stator resistor of balanced three-phase circuits.

(2) Rotor voltage equations [30]:

$$\begin{cases} e_{fd} = p\Psi_{fd} + R_{fd}i_{fd} \\ 0 = p\Psi_{kd} + R_{kd}i_{kd} \\ 0 = p\Psi_{kq} + R_{kq}i_{kq} \end{cases}, \tag{1.65}$$

where e_{fd} is the excitation voltage. Ψ_{fd}, R_{fd}, and i_{fd} are quantities related to the excitation windings. In contrast, Ψ_{kdq}, R_{kdq}, and i_{kdq} refer to the amortisseurs or damping windings.

(3) Stator flux linkage equations [30]:

$$\begin{cases} \Psi_d = -L_d i_d + L_{afd} i_{fd} + L_{akd} i_{kd} \\ \Psi_q = -L_q i_q + L_{akq} i_q \\ \Psi_0 = -L_0 i_0 \end{cases}, \tag{1.66}$$

where L_{dq0} denote the stator self-inductances, while L_{afd} and L_{akdq} represent the mutual inductances between the stator and rotor. Note that the stator flux linkages are intermediate variables in (1.64).

(4) Rotor flux linkage equations [30]:

$$\begin{cases} \Psi_{fd} = L_{ffd} i_{fd} + L_{fkd} i_{kd} - \frac{3}{2} L_{afd} i_d \\ \Psi_{kd} = L_{kfd} i_{fd} + L_{kkd} i_{kd} - \frac{3}{2} L_{akd} i_d, \\ \Psi_{kq} = L_{kkq} i_{kq} - \frac{3}{2} L_{akq} i_q \end{cases} \tag{1.67}$$

where L_{ffd} and L_{kfdq} denote the rotor self-inductances. L_{fkd} and L_{kkd} are the mutual inductances between excitation and damping windings.

(5) Power and torque equations [30]:

$$\begin{aligned} P_t &= \frac{3}{2}(e_d i_d + e_q i_q + 2e_0 i_0) \\ &= \underbrace{\frac{3}{2}(i_d p\Psi_d + i_q p\Psi_q + 2i_0 p\Psi_0)}_{\text{Rate of change of armature magnetic energy}} + \underbrace{\frac{3}{2}\omega_r(i_q\Psi_d - i_d\Psi_q)}_{\text{Power transferred across the air-gap}} \\ &\quad \underbrace{-\frac{3}{2}R_a(i_d^2 + i_q^2 + 2i_0^2)}_{\text{Armature resistance loss}}. \end{aligned} \tag{1.68}$$

$$T_e = \frac{3}{2}\frac{p_f}{2}(\Psi_d i_q - \Psi_q i_d), \tag{1.69}$$

where T_e refers to the electrical torque in Fig. 1.31. The above torque equation is a link between mechanical and electrical models of SGs. In the aforementioned equations, all the inductances are measured or given by SG manufactures. Flux linkages are intermediate variables that can be canceled out. As a result, we control the rotor quantities, such as excitation voltages and mechanical torques, to regulate stator voltages and currents. This is the fundamental principle of SG control.

Per Unit Electrical Models of SGs

Similar to motion equations, we normally reorganize electrical equations of SGs in terms of per unit notations. However, as electrical equations involve more variables, we should select references first. For selection of base quantities, we have multiple options. We choose the L_{ad}-base reciprocal per unit system here [30].

In terms of the voltage base, we select the peak value of rated phase-to-neutral voltages e_{sbase} (V). Similarly, the current base is the peak value of rated line currents i_{sbase} (A). The rated frequency f_{base} (Hz) equals the fundamental frequency f_o.

Next, we derive the rated values of other quantities. $\omega_{base} = 2\pi f_{base}$ (rad/s). As mentioned, the mechanical angular velocity equals the ratio of the electrical angular velocity and the number of pole pairs. Correspondingly, the rated mechanical angular velocity is $\omega_{sm\_base} = 2\omega_{base}/p_f$ (rad/s). The rated stator impedance is calculated as the ratio of the rated voltage and current, i.e., $Z_{sbase} = e_{sbase}/i_{sbase}$ (Ω). Further, the rated stator inductance is $L_{sbase} = Z_{sbase}/\omega_{base}$ (H). The flux-linkage base is calculated by multiplying the rated inductance and current as $\Psi_{sbase} = L_{sbase} i_{sbase} = e_{sbase}/\omega_{base}$ (Wb). Notably, the rated power is derived from the rated voltage and current as $VA_{base} = 3e_{sbase}i_{sbase}/2$ (VA). In addition, the rated torque is $T_{base} = VA_{base}/\omega_{sm\_base}$ $= 3p_f\Psi_{sbase}i_{sbase}/4$ (Nm). For simplification of calculation, the time base is chosen to be $t_{base} = 1/\omega_{base}$ (s).

Subsequently, we derive the per unit inductances. The per unit stator mutual inductances that are introduced by the rotor are expressed as [30]

$$L_{afd\_pu} = \frac{L_{afd}}{L_{sbase}} \cdot \frac{i_{fdbase}}{i_{sbase}},\ L_{akd\_pu} = \frac{L_{akd}}{L_{sbase}} \cdot \frac{i_{kdbase}}{i_{sbase}},\ L_{akq\_pu} = \frac{L_{akq}}{L_{sbase}} \cdot \frac{i_{kqbase}}{i_{sbase}}. \tag{1.70}$$

To eliminate the coefficient 3/2, the per unit rotor mutual inductances due to the stator are derived as [30]

$$L_{fda\_pu} = \frac{3}{2} \cdot \frac{L_{afd}}{L_{fdbase}} \cdot \frac{i_{sbase}}{i_{fdbase}},\ L_{kda\_pu} = \frac{3}{2} \cdot \frac{L_{akd}}{L_{kdbase}} \cdot \frac{i_{sbase}}{i_{kdbase}},$$
$$L_{kqa\_pu} = \frac{3}{2} \cdot \frac{L_{akq}}{L_{kqbase}} \cdot \frac{i_{sbase}}{i_{kqbase}}. \tag{1.71}$$

We derive the per unit rotor mutual inductances among rotor windings as [30]

$$L_{\text{fkd\_pu}} = \frac{L_{\text{fkd}}}{L_{\text{fdbase}}} \cdot \frac{i_{\text{kdbase}}}{i_{\text{fdbase}}},\; L_{\text{kdf\_pu}} = \frac{3}{2} \cdot \frac{L_{\text{fkd}}}{L_{\text{kdbase}}} \cdot \frac{i_{\text{fdase}}}{i_{\text{kdbase}}}. \tag{1.72}$$

Further, we can greatly simplify analysis by carefully selecting the rated rotor currents. In the first step, we equalize the two per unit inductances in (1.72), leading to [30]

$$L_{\text{fkd\_pu}} = L_{\text{kdf\_pu}} \Rightarrow e_{\text{kdbase}} i_{\text{kdbase}} = e_{\text{fdbase}} i_{\text{fdbase}}. \tag{1.73}$$

In the second step, we make the corresponding mutual inductances in (1.70) and (1.71) identical, yielding [30]

$$\begin{aligned} L_{\text{afd\_pu}} &= L_{\text{fda\_pu}} \Rightarrow e_{\text{fdbase}} i_{\text{fdbase}} = \frac{3}{2} e_{\text{sbase}} i_{\text{sbase}} \Rightarrow Z_{\text{fdbase}} \\ &= \frac{VA_{\text{base}}}{i^2_{\text{fdbase}}} \Rightarrow L_{\text{fdbase}} = \frac{Z_{\text{fdbase}}}{\omega^2_{\text{base}}}, \end{aligned} \tag{1.74}$$

$$\begin{aligned} L_{\text{akd\_pu}} &= L_{\text{kda\_pu}} \Rightarrow e_{\text{kdbase}} i_{\text{kdbase}} = \frac{3}{2} e_{\text{sbase}} i_{\text{sbase}} \Rightarrow Z_{\text{kdbase}} \\ &= \frac{VA_{\text{base}}}{i^2_{\text{kdbase}}} \Rightarrow L_{\text{kdbase}} = \frac{Z_{\text{kdbase}}}{\omega^2_{\text{base}}}, \end{aligned} \tag{1.75}$$

$$\begin{aligned} L_{\text{akq\_pu}} &= L_{\text{kqa\_pu}} \Rightarrow e_{\text{kqbase}} i_{\text{kqbase}} = \frac{3}{2} e_{\text{sbase}} i_{\text{sbase}} \Rightarrow Z_{\text{kqbase}} \\ &= \frac{VA_{\text{base}}}{i^2_{\text{kqbase}}} \Rightarrow L_{\text{kqbase}} = \frac{Z_{\text{kqbase}}}{\omega^2_{\text{base}}}. \end{aligned} \tag{1.76}$$

So far, we have fixed all the rated rotor currents. Considering (1.73)–(1.75), we have [30]

$$L_{\text{ad\_pu}} = L_{\text{afd\_pu}} = L_{\text{fda\_pu}} = L_{\text{akd\_pu}} = L_{\text{kda\_pu}}. \tag{1.77}$$

Moreover, (1.76) indicates that

$$L_{\text{aq\_pu}} = L_{\text{akq\_pu}} = L_{\text{kqa\_pu}}. \tag{1.78}$$

On top of that, (1.73) tells that

$$L_{\text{fkd\_pu}} = L_{\text{kdf\_pu}}. \tag{1.79}$$

The respective base currents are expressed as

$$i_{\text{fdbase}} = i_{\text{sbase}} \frac{L_{\text{ad}}}{L_{\text{afd}}},\ i_{\text{kdbase}} = i_{\text{sbase}} \frac{L_{\text{ad}}}{L_{\text{akd}}},\ i_{\text{kqbase}} = i_{\text{sbase}} \frac{L_{\text{aq}}}{L_{\text{akq}}}. \tag{1.80}$$

After knowing the base quantities and per unit inductances, we can reorganize the electrical equations of SGs in per unit forms as

(1) Per unit stator voltage equations [30]:

$$\begin{cases} e_{\text{d\_pu}} = p_{\text{pu}} \Psi_{\text{d\_pu}} - \Psi_{\text{q\_pu}} \omega_{\text{r\_pu}} - R_{\text{a\_pu}} i_{\text{d\_pu}} \\ e_{\text{q\_pu}} = p_{\text{pu}} \Psi_{\text{q\_pu}} + \Psi_{\text{d\_pu}} \omega_{\text{r\_pu}} - R_{\text{a\_pu}} i_{\text{q\_pu}}, \\ e_{\text{0\_pu}} = p_{\text{pu}} \Psi_{\text{0\_pu}} - R_{\text{a\_pu}} i_{\text{0\_pu}} \end{cases} \tag{1.81}$$

where p_{pu} refers to the per unit differential operator, i.e., $p_{\text{pu}} = 1/\omega_{\text{base}} \times \text{d}()/\text{d}t$.

(2) Per unit rotor voltage equations [30]:

$$\begin{cases} e_{\text{fd\_pu}} = p_{\text{pu}} \Psi_{\text{fd\_pu}} + R_{\text{fd\_pu}} i_{\text{fd\_pu}} \\ 0 = p_{\text{pu}} \Psi_{\text{kd\_pu}} + R_{\text{kd\_pu}} i_{\text{kd\_pu}} \\ 0 = p_{\text{pu}} \Psi_{\text{kq\_pu}} + R_{\text{kq\_pu}} i_{\text{kq\_pu}} \end{cases}. \tag{1.82}$$

(3) Per unit stator flux linkage equations [30]:

$$\begin{cases} \Psi_{\text{d\_pu}} = -L_{\text{d\_pu}} i_{\text{d\_pu}} + L_{\text{ad\_pu}} i_{\text{fd\_pu}} + L_{\text{ad\_pu}} i_{\text{kd\_pu}} \\ \Psi_{\text{q\_pu}} = -L_{\text{q\_pu}} i_{\text{q\_pu}} + L_{\text{aq\_pu}} i_{\text{kq\_pu}} \\ \Psi_{\text{0\_pu}} = -L_{\text{0\_pu}} i_{\text{0\_pu}} \end{cases}, \tag{1.83}$$

where $L_{\text{d\_pu}} = L_{\text{ad\_pu}} + L_{\text{l\_pu}}$ and $L_{\text{q\_pu}} = L_{\text{aq\_pu}} + L_{\text{l\_pu}}$.

(4) Per unit rotor flux linkage equations [30]:

$$\begin{cases} \Psi_{\text{fd\_pu}} = L_{\text{ffd\_pu}} i_{\text{fd\_pu}} + L_{\text{fkd\_pu}} i_{\text{kd\_pu}} - L_{\text{ad\_pu}} i_{\text{d\_pu}} \\ \Psi_{\text{kd\_pu}} = L_{\text{fkd\_pu}} i_{\text{fd\_pu}} + L_{\text{kkd\_pu}} i_{\text{kd\_pu}} - L_{\text{ad\_pu}} i_{\text{d\_pu}}. \\ \Psi_{\text{kq\_pu}} = L_{\text{kkq\_pu}} i_{\text{kq\_pu}} - L_{\text{aq\_pu}} i_{\text{q\_pu}} \end{cases} \tag{1.84}$$

(5) Per unit power and torque equations [30]:

$$P_{\text{t\_pu}} = e_{\text{d\_pu}} i_{\text{d\_pu}} + e_{\text{q\_pu}} i_{\text{q\_pu}} + 2 e_{\text{0\_pu}} i_{\text{0\_pu}} \text{ and} \tag{1.85}$$

$$T_{\text{e\_pu}} = \Psi_{\text{d\_pu}} i_{\text{q\_pu}} - \Psi_{\text{q\_pu}} i_{\text{d\_pu}}, \tag{1.86}$$

where the per unit torque $T_{\text{e\_pu}}$ appears in the swing equation (1.46)–(1.48). The per unit electrical equations of SGs accurately describe the dynamics behaviors of SGs.

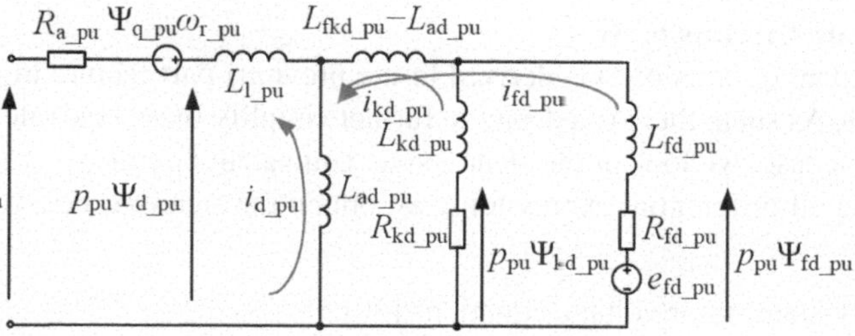

Fig. 1.34 *D*-axis equivalent circuit of synchronous generators

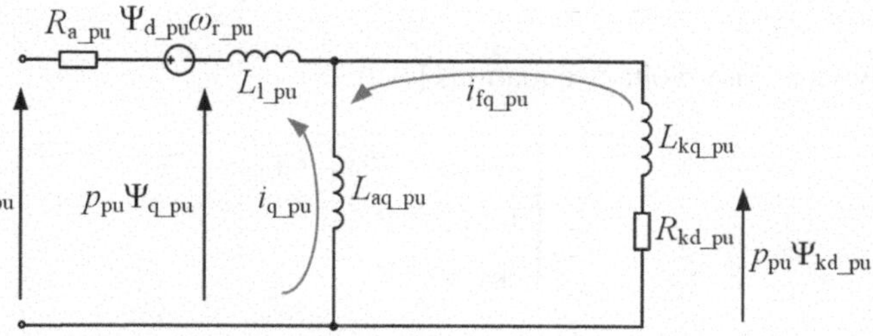

Fig. 1.35 *Q*-axis equivalent circuit of synchronous generators

Fig. 1.36 0-axis equivalent circuit of synchronous generators

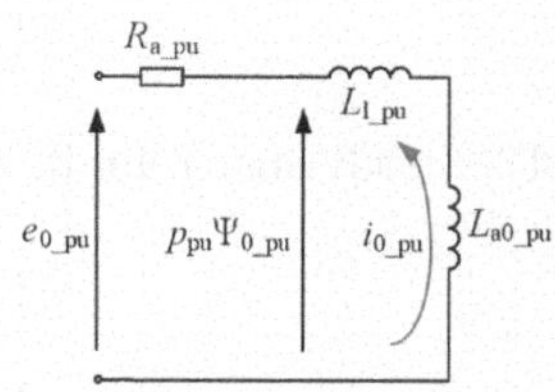

Equivalent Circuits of SGs

With the per unit voltage and flux linkage equations, we can further draw the equivalent circuits of SGs for better illustration. It should be kept in mind that amortisseurs can be modeled either with one or more than one *q*-axis equations, leading to different equivalent circuits [30]. For simplicity, we consider the case of one equation.

Figure 1.34 demonstrates the *d*-axis equivalent circuit of SGs, where the *d*-axis voltage, current, and flux linkage quantities are included. $L_{\mathrm{fd\_pu}}$ denotes the inductance difference between $L_{\mathrm{ffd\_pu}}$ and $L_{\mathrm{fkd\_pu}}$, namely, $L_{\mathrm{fd\_pu}} = L_{\mathrm{ffd\_pu}} - L_{\mathrm{fkd\_pu}}$. Moreover, $L_{\mathrm{kd\_pu}}$ is defined as $(L_{\mathrm{kkd\_pu}} - L_{\mathrm{fkd\_pu}})$. It should be highlighted that the excitation voltage $e_{\mathrm{fd\_pu}}$ is controllable through external voltage supplies.

Figure 1.35 shows the *q*-axis equivalent circuit of SGs, where the *q*-axis voltage, current, and flux linkage quantities are incorporated. Notably, $L_{\mathrm{kq\_pu}}$ is equal to $(L_{\mathrm{kkq\_pu}} - L_{\mathrm{aq\_pu}})$. Different from the *d*-axis equivalent circuit, the *q*-axis equivalent circuit contains no excitation circuit.

Figure 1.36 shows the 0-axis equivalent circuit of SGs, which can be ignored for three-phase balanced circuits.

Steady-State Circuits of SGs

The equivalent circuits of SGs derived in the previous part feature high orders and complexity. As such, there is a desire to further simplify these equivalent circuits. In the simplest case, we look at the steady-state equivalent circuits.

Ignoring all differential expressions, we obtain the steady-state SG equations as follows.

(1) Steady-state stator voltage equations [30]:

$$\begin{cases} e_{\mathrm{d\_pu}} = -\Psi_{\mathrm{q\_pu}}\omega_{\mathrm{r\_pu}} - R_{\mathrm{a\_pu}}i_{\mathrm{d\_pu}} \\ e_{\mathrm{q\_pu}} = \Psi_{\mathrm{d\_pu}}\omega_{\mathrm{r\_pu}} - R_{\mathrm{a\_pu}}i_{\mathrm{q\_pu}} \end{cases}. \tag{1.87}$$

(2) Steady-state rotor voltage equations [30]:

$$\begin{cases} e_{\mathrm{fd\_pu}} = R_{\mathrm{fd\_pu}}i_{\mathrm{fd\_pu}} \\ i_{\mathrm{kd\_pu}} = 0 \\ i_{\mathrm{kq\_pu}} = 0 \end{cases}. \tag{1.88}$$

(3) Steady-state stator flux linkage equations [30]:

$$\begin{cases} \Psi_{\mathrm{d\_pu}} = -L_{\mathrm{d\_pu}}i_{\mathrm{d\_pu}} + L_{\mathrm{ad\_pu}}i_{\mathrm{fd\_pu}} \\ \Psi_{\mathrm{q\_pu}} = -L_{\mathrm{q\_pu}}i_{\mathrm{q\_pu}} \end{cases}. \tag{1.89}$$

(4) Steady-state rotor flux linkage equations [30]:

$$\begin{cases} \Psi_{\mathrm{fd\_pu}} = L_{\mathrm{ffd\_pu}}i_{\mathrm{fd\_pu}} - L_{\mathrm{ad\_pu}}i_{\mathrm{d\_pu}} \\ \Psi_{\mathrm{kd\_pu}} = L_{\mathrm{fkd\_pu}}i_{\mathrm{fd\_pu}} - L_{\mathrm{ad\_pu}}i_{\mathrm{d\_pu}}. \\ \Psi_{\mathrm{kq\_pu}} = -L_{\mathrm{aq\_pu}}i_{\mathrm{q\_pu}} \end{cases} \tag{1.90}$$

The power and torque equations remain unchanged in steady state, as described by (1.85) and (1.86), respectively. Substitution of the steady-state stator flux linkage equations (1.89) into the steady-state stator voltage equations (1.87), the latter becomes [30]

$$\begin{cases} e_{\mathrm{d\_pu}} = X_{\mathrm{q\_pu}}i_{\mathrm{q\_pu}} - R_{\mathrm{a\_pu}}i_{\mathrm{d\_pu}} \\ e_{\mathrm{q\_pu}} = -X_{\mathrm{d\_pu}}i_{\mathrm{d\_pu}} + X_{\mathrm{ad\_pu}}i_{\mathrm{fd\_pu}} - R_{\mathrm{a\_pu}}i_{\mathrm{q\_pu}} \end{cases}, \tag{1.91}$$

where $X_{\mathrm{d\_pu}}$ and $X_{\mathrm{q\_pu}}$ denote the per unit *d*- and *q*-axis reactance, respectively. $X_{\mathrm{ad\_pu}} = L_{\mathrm{ad\_pu}}\omega_{\mathrm{r\_pu}}$ stands for the mutual reactance between stator and rotor. We can derive the per unit excitation current $i_{\mathrm{fd\_pu}}$ from (1.88), and then solve stator currents and voltages based on the V–I relationship of loads. Alternatively, given that stator currents and voltages are known, we derive the excitation current from (1.89) as [30]

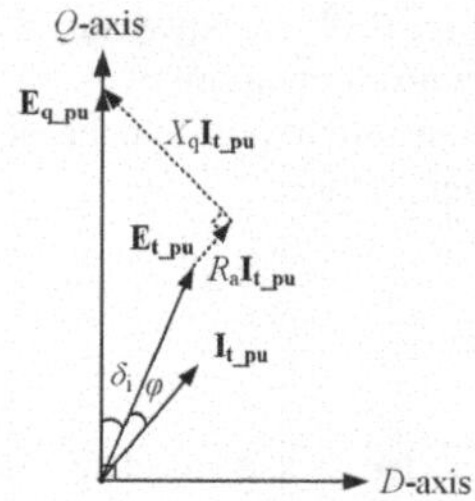

Fig. 1.37 Phasor diagram of the steady-state model of synchronous generators

$$i_{fd\_pu} = \frac{e_{q\_pu} + R_{a\_pu} i_{q\_pu} + X_{c\_pu} i_{d\_pu}}{X_{ad\_pu}}, \tag{1.92}$$

which gives a reference excitation current that facilitates SG voltage regulation.

The terminal voltage and current relationships described by (1.91) enables steady-state circuit analysis in terms of state equations. To further analyze steady-state relationships through the classic control theory, we express electrical quantities as complex phasors. To be specific, the stator-voltage phasor is

$$\mathbf{E}_{t\_pu} = e_{d\_pu} + j e_{q\_pu} = E_m \angle (\frac{\pi}{2} - \delta_i), \tag{1.93}$$

where the phase angle is intentionally expressed as a complimentary form for further analysis. Assuming that inductive loads cause the stator-current phasor to lag the stator-voltage phasor by an angle of φ, that is,

$$\mathbf{I}_{t\_pu} = i_{d\_pu} + j i_{q\_pu} = I_m \angle (\frac{\pi}{2} - \delta_i - \varphi). \tag{1.94}$$

Further, we define an internal-voltage phasor as

$$\mathbf{E}_{q\_pu} = \mathbf{E}_{t\_pu} + (R_{a\_pu} + j X_{q\_pu}) \mathbf{I}_{t\_pu}. \tag{1.95}$$

Substitution of (1.91) into (1.93) and then the result, together with (1.94), into (1.95), it yields [30]

$$\mathbf{E}_{q\_pu} = j(X_{q\_pu} - X_{d\_pu}) i_{d\_pu} + j X_{ad\_pu} i_{fd\_pu}, \tag{1.96}$$

which is in alignment with the q-axis, and hence with the subscript q. For demonstration, Fig. 1.37 shows a phasor diagram of the steady-state model of SGs. Clearly, the internal-voltage phasor is located in the q-axis. The phase angle δ_i stands for the angle difference between the terminal and stator voltage phasors, which is defined as the load angle.

Under no load conditions, the stator currents become zero, resulting in

$$\mathbf{E}_{t\_pu} = \mathbf{E}_{q\_pu} = j X_{ad\_pu} i_{fd\_pu}. \tag{1.97}$$

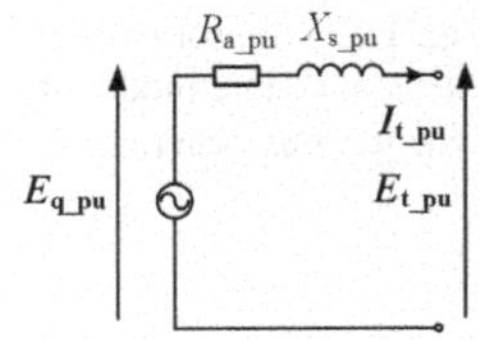

Fig. 1.38 Steady-state equivalent circuit of non-salient pole synchronous generators

In this case, the stator voltage amplitude can be adjusted in proportion to the excitation current i_{fd_pu}. Moreover, the voltage phasor always leads the excitation current phasor by an angle of 90°.

For non-salient pole SGs, we can assume that *d*- and *q*-axis impedances are identical, thereby leading to

$$X_{s\_pu} = X_{d\_pu} = X_{q\_pu}, \tag{1.98}$$

whose reciprocal is defined as the *short-circuit ratio* (SCR):

$$\text{SCR} = 1/X_{s\_pu}. \tag{1.99}$$

Substitution of (1.98) into (1.96), it yields

$$\mathbf{E_{q\_pu}} = jX_{ad\_pu}i_{fd\_pu}, \tag{1.100}$$

whose amplitude is independent of stator currents. Finally, Fig. 1.38 presents the steady-state equivalent circuit of non-salient pole SGs, which is very simple.

The output complex power of SGs is calculated from Fig. 1.38 as [30]

$$\begin{aligned} \mathbf{S_{t\_pu}} &= \mathbf{E_{t\_pu}}\mathbf{I^*_{t\_pu}} = P_{t\_pu} + jQ_{t\_pu} \\ &= (e_{d\_pu}i_{d\_pu} + e_{q\_pu}i_{q\_pu}) + j(e_{q\_pu}i_{d\_pu} - e_{d\_pu}i_{q\_pu}), \end{aligned} \tag{1.101}$$

where the asterisk notation represents the complex conjugate operation. Additionally, we derive the per unit electrical torque from (1.86), (1.87), and (1.89) as [30]

$$T_{e\_pu} = P_{t\_pu} + R_a\left|\mathbf{I_{t\_pu}}\right|^2. \tag{1.102}$$

Simplified Electrical Circuits of SGs

Next, we simplify the equivalent circuits of SGs. The simplification is under three assumptions [30]:

(1) Ignorance of the stator transformer voltage terms, i.e., the time-variant stator flux linkages $p_{pu}\mathbf{\Psi_d}$ and $p_{pu}\mathbf{\Psi_q}$.
(2) Constant speed, i.e., $\omega_{r\_pu} = 1.0$.
(3) Removal of amortisseurs.

As a result, the simplified equations of SGs are

(1) Simplified per unit stator voltage equations:

$$\begin{cases} e_{\mathrm{d\_pu}} = -\Psi_{\mathrm{q\_pu}}\omega_{\mathrm{r\_pu}} - R_{\mathrm{a\_pu}}i_{\mathrm{d\_pu}} \\ e_{\mathrm{q\_pu}} = \Psi_{\mathrm{d\_pu}}\omega_{\mathrm{r\_pu}} - R_{\mathrm{a\_pu}}i_{\mathrm{q\_pu}} \end{cases}. \tag{1.103}$$

(2) Simplified per unit rotor voltage equation:

$$e_{\mathrm{fd\_pu}} = p_{\mathrm{pu}}\Psi_{\mathrm{fd\_pu}} + R_{\mathrm{fd\_pu}}i_{\mathrm{fd\_pu}}. \tag{1.104}$$

(3) Simplified per unit stator flux linkage equations:

$$\begin{cases} \Psi_{\mathrm{d\_pu}} = -L_{\mathrm{d\_pu}}i_{\mathrm{d\_pu}} + L_{\mathrm{ad\_pu}}i_{\mathrm{fd\_pu}} \\ \Psi_{\mathrm{q\_pu}} = -L_{\mathrm{q\_pu}}i_{\mathrm{q\_pu}} \end{cases}. \tag{1.105}$$

(4) Simplified per unit rotor flux linkage equation:

$$\Psi_{\mathrm{fd\_pu}} = L_{\mathrm{ffd\_pu}}i_{\mathrm{fd\_pu}} - L_{\mathrm{ad\_pu}}i_{\mathrm{d\_pu}}. \tag{1.106}$$

Alternatively, we can reorganize the above relationships into [30]

$$\begin{cases} \Psi_{\mathrm{d\_pu}} = -L_{\mathrm{d\_pu}}i_{\mathrm{d\_pu}} + E_{\mathrm{I\_pu}} \\ \Psi_{\mathrm{q\_pu}} = -L_{\mathrm{q\_pu}}i_{\mathrm{q\_pu}} \\ E'_{\mathrm{q\_pu}} = E_{\mathrm{I\_pu}} - (L_{\mathrm{d\_pu}} - L'_{\mathrm{d\_pu}})i_{\mathrm{d\_pu}} \\ p_{\mathrm{pu}}E'_{\mathrm{q\_pu}} = \dfrac{E_{\mathrm{fd\_pu}} - E_{\mathrm{I\_pu}}}{T'_{\mathrm{d0}}} \end{cases}, \tag{1.107}$$

where

$$\begin{aligned} E_{\mathrm{I\_pu}} &= L_{\mathrm{ad\_pu}}i_{\mathrm{fd\_pu}} \\ E'_{\mathrm{q\_pu}} &= \frac{L_{\mathrm{ad\_pu}}}{L_{\mathrm{ffd\_pu}}}\Psi_{\mathrm{fd\_pu}} = -\frac{L^2_{\mathrm{ad\_pu}}}{L_{\mathrm{ffd\_pu}}}i_{\mathrm{d\_pu}} + E_{\mathrm{I\_pu}} \\ E_{\mathrm{fd\_pu}} &= \frac{L_{\mathrm{ad\_pu}}}{R_{\mathrm{fd\_pu}}}e_{\mathrm{fd\_pu}} \end{aligned}. \tag{1.108}$$

In addition, $L_{\mathrm{d\_pu}}'$ (or $X_{\mathrm{d\_pu}}'$) and T_{d0}' can be obtained from SG tests. The relationships among the voltage phasors are [30]

$$\begin{aligned} \mathbf{E}_{\mathbf{q\_pu}} &= \mathbf{E}_{\mathbf{t\_pu}} + (R_{\mathrm{a\_pu}} + jX_{\mathrm{q\_pu}})\mathbf{I}_{\mathbf{t\_pu}} \\ \mathbf{E}_{\mathbf{I\_pu}} &= \mathbf{E}_{\mathbf{q\_pu}} + j(X_{\mathrm{d\_pu}} - X_{\mathrm{q\_pu}})\mathrm{Re}(\mathbf{I}_{\mathbf{t\_pu}}) \\ \mathbf{E}'_{\mathbf{pu}} &= \mathbf{E}_{\mathbf{t\_pu}} + (R_{\mathrm{a\_pu}} + jX'_{\mathrm{d\_pu}})\mathbf{I}_{\mathbf{t\_pu}} \\ E'_{\mathrm{q\_pu}} &= \mathrm{Im}(\mathbf{E}'_{\mathbf{pu}}) \end{aligned}, \tag{1.109}$$

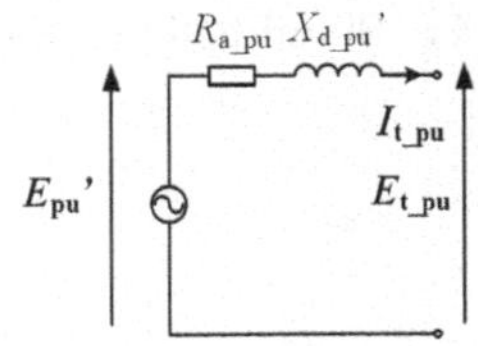

Fig. 1.39 Simplified dynamic equivalent circuit of non-salient pole synchronous generators

where Re() and Im() refer to the functions of extracting real and imaginary parts, respectively.

For non-salient SGs, we have [30]

$$\begin{aligned}
&X_{\mathrm{d_pu}} = X_{\mathrm{q_pu}} \\
&\mathbf{E}_{\mathbf{I_pu}} = \mathbf{E}_{\mathbf{q_pu}} = \mathbf{E}_{\mathbf{t_pu}} + (R_{\mathrm{a_pu}} + jX_{\mathrm{q_pu}})\mathbf{I}_{\mathbf{t_pu}} \\
&\mathbf{E}'_{\mathbf{pu}} = \mathbf{E}_{\mathbf{t_pu}} + (R_{\mathrm{a_pu}} + jX'_{\mathrm{d_pu}})\mathbf{I}_{\mathbf{t_pu}} = jE'_{\mathrm{q_pu}} = j\left(-\frac{L^2_{\mathrm{ad_pu}}}{L_{\mathrm{ffd_pu}}} i_{\mathrm{d_pu}} + E_{\mathrm{I_pu}}\right)
\end{aligned} \tag{1.110}$$

The corresponding equivalent circuit is shown in Fig. 1.39. Although the resultant equivalent circuit is very simple, it should be kept in mind that the assumptions mentioned before may not be valid in practice. In that case, the original equivalent circuits (see Figs. 1.34 and 1.35) should be used.

We have detailed the models of SGs, as they are the most important power generators in bulk power systems. Although power converters tend to replace SGs in more-electronics power systems (as will be introduced), they normally do not possess all useful features of SGs, thereby leading to stability and power quality problems. The key of addressing such problems lies in the emulation of SG behaviors to some extents. However, as presented, the models of SGs can be quite complicated. As such, the identification of decisive parts of SG models becomes very important. In addition, the retirement of SGs will not be achieved immediately. Therefore, we expect power converters and SGs to coexist in the near future. That is why we elaborate SG models. After modeling SGs, we continue to explore the models of other parts in the process of power generation.

Models of Turbine and Governing Systems

Turbines and speed governors determine the mechanical torque of SGs, which further controls the motion of SGs and active power. Therefore, we should explore their models.

First, let us look at the turbines of SGs, which are supplied by heat sources, such as coal, oil, and gas, etc. We regulate the control valves of turbines to adjust the mechanical torque of SGs. For simplification, we ignore the complicated mechanical structures of turbines and present a classic model of fossil-fueled single reheat turbines as [30]

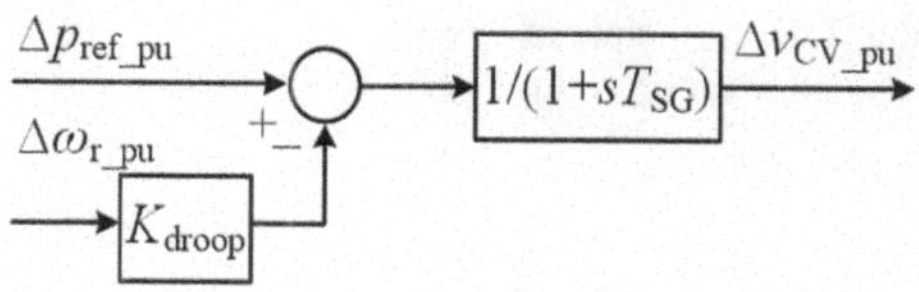

Fig. 1.40 Linearized model of speed governors

$$G_{\text{Turbine}}(s) = \frac{\Delta T_{\text{m\_pu}}(s)}{\Delta v_{\text{CV\_pu}}(s)} = \frac{1 + F_{\text{HP}} T_{\text{RH}} s}{(1 + T_{\text{CH}} s)(1 + T_{\text{RH}} s)}, \tag{1.111}$$

where T_{CH} stands for the time constant of the response of steam flow to a change in control valve opening [30]. T_{RH} represents the time constant associated with the reheater [30]. F_{HP} refers to a power fraction factor [30]. The model in (1.111) assumes that control valves are linear. In addition, note that (1.111) is a per unit transfer function, whose power base is the maximum turbine power, which can be different from that of SGs. Typical values of coefficients are listed as follows. T_{CH} = 0.3 s, T_{RH} = 7.0 s, and F_{HP} = 0.3 s.

Next, we move on to the model of speed governors. Speed governors are controllers of SGs for speed regulation. The two inputs of speed governors are the load reference and perturbed rotor angular velocity, while the single output controls the valves of turbines. Figure 1.40 shows a linearized model of speed governors, where $\Delta p_{\text{ref\_pu}}$ denotes the perturbed load reference, $\Delta\omega_{\text{r\_pu}}$ refers to the perturbed rotor angular velocity (which equals the perturbed frequency in per unit), and $\Delta v_{\text{CV\_pu}}$ stands for the input of turbines [see (1.111)]. As observed, we model speed governors as a first-order inertial element:

$$G_{\text{Governor}}(s) = \frac{1}{1 + T_{\text{SG}} s}, \tag{1.112}$$

where T_{SG} is a lumped time constant of speed governors [30]. Generally, T_{SG} ranges from 0.1 to 0.3 s. It should be mentioned that time delays of speed governors are mainly due to mechanical parts.

Special attention should be paid to the droop coefficient K_{droop}. Through frequency droop, SGs automatically share active power according to their respective power ratings. We observe from Fig. 1.40 that speed governors link their reference active power and the rotor speed through K_{droop}. Referring to Fig. 1.41, the increase of frequency will reduce the active power of SGs, thereby helping active power balance and stabilizing system frequency. For a larger K_{droop}, the same change of frequency $\Delta\omega_{\text{r\_pu}}$ will cause a more pronounced power drop. As such, power converters can exploit different droop coefficients according to their respective power ratings to share power accordingly, as power converters see the same frequency in steady state. We will further discuss the effects of droop control later.

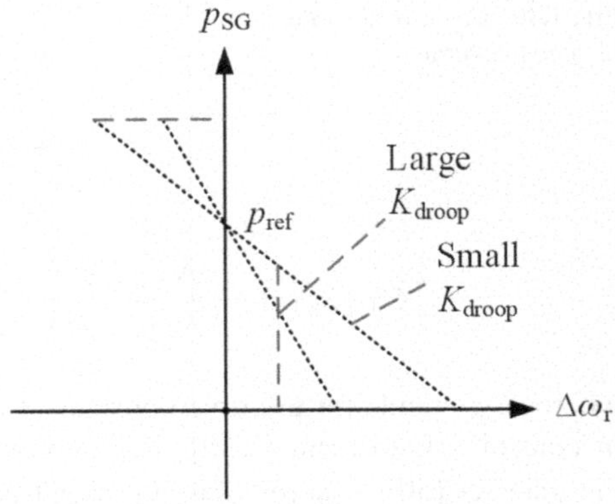

Fig. 1.41 Principle of frequency droop control

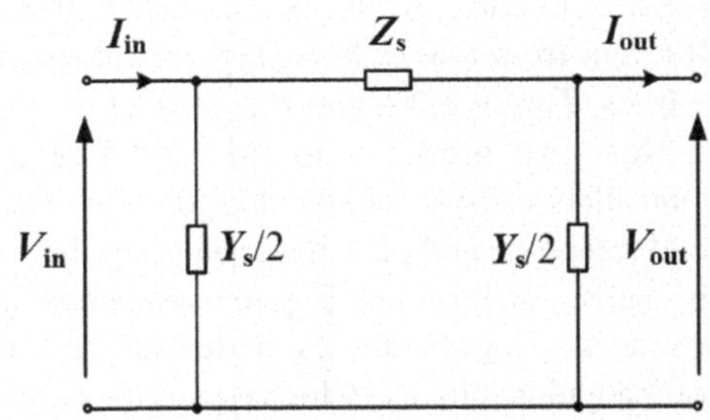

Fig. 1.42 Lumped equivalent circuit of transmission lines

Model of Transmission Lines

We continue to model transmission lines. For simplicity, we focus on the most relevant models.

Transmission lines are generally divided into overhead and cable lines. In comparison, overhead lines are cheaper than cables by a factor of more than 10. Nevertheless, cables feature better protection and visualization. In contrast, cables have larger equivalent capacitances. Although transmission lines are essentially circuits with distributed parameters, we prefer to model them as lumped equivalent circuits for ease of analysis. For example, Fig. 1.42 shows a lumped equivalent circuit of transmission lines. The relevant impedances are derived as [30]

$$\mathbf{Z}_s = \mathbf{Z}_C \sinh(\gamma l) \text{ and} \tag{1.113}$$

$$\frac{\mathbf{Y}_s}{2} = \frac{1}{\mathbf{Z}_C} \tanh(\gamma l), \tag{1.114}$$

where l stands for the length of transmission lines. $\mathbf{Z}_C$ and γ are the characteristic impedance and propagation coefficient, respectively. They are expressed as [30]

$$\mathbf{Z}_C = \sqrt{\frac{L_{s0}}{C_{s0}}}(1 - j\frac{R_{s0}}{2\omega_0 L_{s0}}) \text{ and} \tag{1.115}$$

$$\gamma = \sqrt{L_{s0}C_{s0}}(\frac{R_{s0}}{2L_{s0}} + j\omega_0), \tag{1.116}$$

where R_{s0}, L_{s0}, and C_{s0} represent the equivalent series resistance, series inductance, and parallel capacitance of transmission lines per length. As long as the geometrical shapes of transmission lines are known, we can calculate or obtain R_{s0}, L_{s0}, and C_{s0} from datasheets.

For highly inductive lines, we can further simplify the model of transmission lines as an inductor or a serial *RL* circuit [37]. By doing so, system analysis and design can greatly be simplified.

Model of Loads

The models of loads vary significantly and can be very complicated depend on the nature and amount of loads. In this part, we concentrate on the most easy-to-use and standard load model. It should be pointed out that inductor motors stand for the majority of loads.

The most straight forward approach of load modeling is performed in terms of active and reactive power consumption. In steady-state, active power and reactive power are functions of voltage amplitude and frequency, i.e.[30],

$$P_{load} = P_{load0}(V_{load\_pu})^{K_{load\_pv}}(1 + K_{load\_pf}\Delta f_{r\_pu}) \text{ and} \tag{1.117}$$

$$Q_{load} = Q_{load0}(V_{load\_pu})^{K_{load\_qv}}(1 + K_{load\_qf}\Delta f_{r\_pu}), \tag{1.118}$$

where P_{load0} and Q_{load0} refer to the rated active power and reactive power of loads, respectively. K_{load_pv} and K_{load_qv} represent the exponential voltage load coefficients of active and reactive power, respectively. Notably, we only consider the linear parts of frequency terms. Correspondingly, K_{load_pf} and K_{load_qf} denote the frequency load coefficients of active and reactive power, respectively.

Typical values of K_{load_pv} and K_{load_qv} are 0, 1, or 2, representing constant power, constant current, or constant impedance loads, respectively. For common loads, K_{load_pv} ranges from 0.5 to 1.8, while K_{load_qv} is typically from 1.5 to 6.0 [30]. K_{load_pv} often introduces nonlinearity due to magnetic saturation.

K_{load_pf} ranges from 0 to 3.0, and K_{load_qf} ranges from -2.0 to 0 [30]. Therefore, loads play an important in terms of frequency and power regulation. The steady-state load models described by (1.117) and (1.118) are sufficient in most cases.

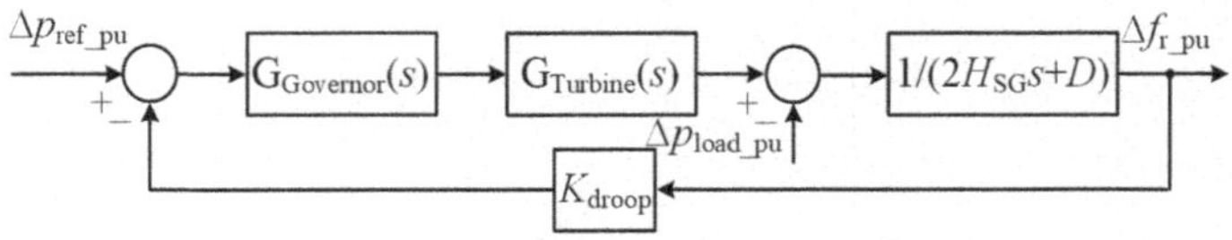

Fig. 1.43 Model of single-islanded synchronous generators with frequency regulation

Model of Single-Islanded SGs with Frequency Regulation

Considering the models of SGs, turbines, speed governors, and loads, we can build an overall model of frequency control frameworks for single-islanded SGs, as depicted in Fig. 1.43, where the perturbed per unit frequency Δf_{r_pu} equals the perturbed per unit angular frequency $\Delta\omega_{r\_pu}$. D stands for the lumped damping factor contributed by SGs [through K_d in (1.47)] and loads [through K_{load_pf} in (1.117)]. Δp_{load_pu} denotes the demanded power. We will use this model in Chap. 4. In addition, grid-tied conditions are excluded in this chapter and will be discussed later.

We have covered the subtopics of power systems, particularly for the models regarding power generation, which are closely related to power quality and stability of more-electronics power systems. Before ending this section, let us look at some power system examples.

1.3.4 Typical and Emerging Power Systems

This subsection presents some typical power systems. On top of that, we will take a view on emerging power systems.

1.3.4.1 Typical Power Systems

Power systems are essential in every country. To list a few examples, we present the major operators, rated frequencies, and rated voltage amplitudes of typical power systems or utilities in Table 1.5 [38].

As shown, most of the utilities share a rated frequency of 50 Hz, except for north America and part of Japan, which use 60 Hz. Additionally, rated voltage amplitudes range from 100 to 230 V.

1.3.4.2 Emerging Power Systems

Microgrids and *smart grids* are emerging power systems. In the past decade, many novel ideas and concepts related to them have been translated into practical projects and products.

Table 1.5 Information of typical power systems

Regions	Operators	Rated frequencies (Hz)	Rated voltage amplitudes (V)
Australia	Australian Energy Market Operator	50	230
China	State Grid	50	220
Europe	European Network of Transmission System Operators	50	230
Japan	National Energy Grid	50/60	100
North America	California ISO, Electric Reliability Council of Texas, etc	60	120
UK	National Grid	50	230

A microgrid is a combination of generators, loads, and often energy storage in a localized fashion. Microgrids may either operate autonomously or synchronize with the mains grid [39]. Notably, *distributed generators* (DGs) or *distributed energy resources* (DERs) are generally employed in microgrids to enable plug-and-play operation. As stated about power system features, microgrids can also be classified into ac, dc, and hybrid ac/dc microgrids [40, 41]. In terms of applications, microgrids can appear in campuses, communities, vehicles, ships, and aircrafts, etc. The modern society witnesses increasing microgrid demonstrations. It should be highlighted that microgrids can be typical more-electronics power systems.

Smart grids have various definitions. Simply put, a smart grid refers to a power grid with widespread use of communication and digital control. Normally, smart grids normally feature a high-penetration level of renewable energy resources, such as PV and wind. Besides, salient features of smart grids include self-restoration, smart meters, and high reliability. Also, the transition from conventional grids to future smart grids should not be an entire replacement but retrofitting [42]. Due to ambiguous definitions of smart grids, we can hardly show any real smart grid example. However, it is certain that modern power grids will eventually develop into smart grids.

1.4 More-Electronics Power Systems

After introducing power electronics and power systems, we proceed to more-electronics power systems, which are essentially combinations of power electronics and power systems. This section will introduce the concept of more-electronics power systems, followed by their key features.

1.4.1 What Are More-Electronics Power Systems?

The marriage of power electronics and power systems gives rise to *more-electronics power systems*. Overall, more-electronics power systems are the power girds where most grid interfaces are power converters. Generally, the percentage of power electronic interfaces should exceed 50%. More-electronics power systems will become *power-electronics-based power systems* if all interfaces are power converters [43].

In bulk power systems, power converters are increasingly employed as grid interfaces in replacement of conventional SGs. Typical driving applications behind such a trend include renewable generation, energy storage, and electric vehicles. To illustrate, Figs. 1.44 and 1.45 present the global capacity and annual additions from 2006 to 2016 of solar PV and wind power, respectively [44].

Clearly, the installation capacity of renewable energies continues to increase, particularly for solar PV, whose annual addition even outweighs that of wind power

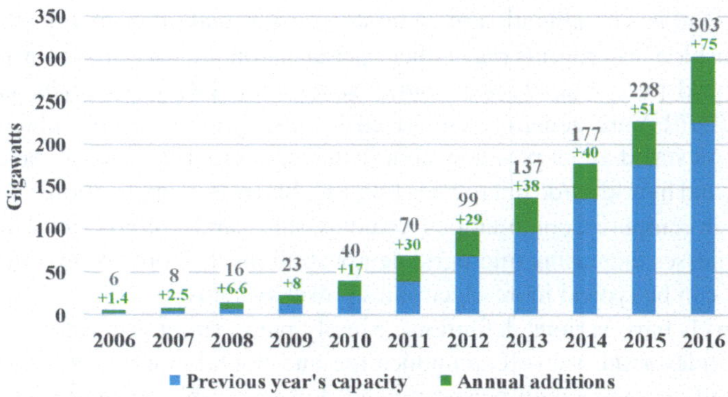

Fig. 1.44 Solar PV global capacity and annual additions

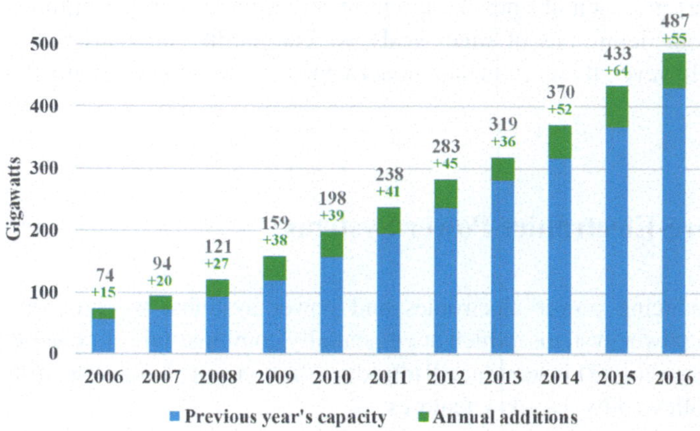

Fig. 1.45 Wind power global capacity and annual additions

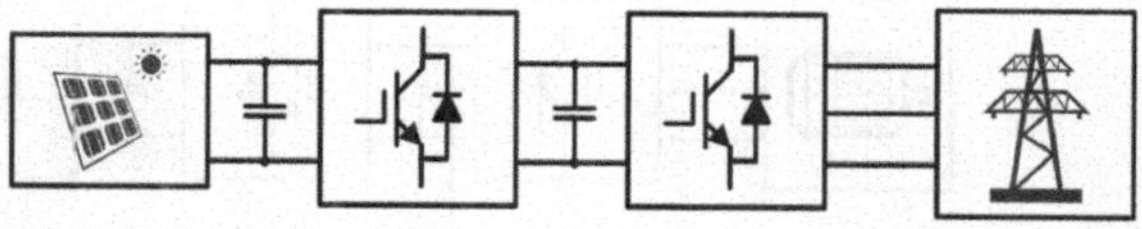

Fig. 1.46 Grid-tied architecture of PV generation systems

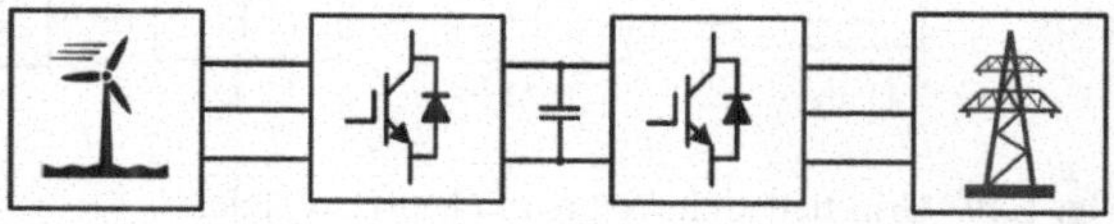

Fig. 1.47 Grid-tied architecture of wind generation systems

in 2016. Renewable energy resources (RESs) are generally coupled to power grids through grid-tied power converters. Figures 1.46 and 1.47 show standard grid-tied architectures of PV and wind generation systems, respectively. It is noticed that PV generation systems achieve grid connection through two power conversion stages—one dc–dc stage and one dc–ac stage. The first dc–dc stage boosts voltages from PV panels into higher dc bus voltages, which are inverted into ac grid voltages through the second dc–ac stage. Different from PV panels, wind turbines output ac voltages, thus necessitating one ac–dc stage.

To get rid of power intermittency (due to weather conditions) and fluctuations from RESs, a large amount of energy storage systems (ESSs) is deployed in modern power systems. For example, Fig. 1.48 shows a grid-tied architecture of battery storage systems. Notably, power converters in Fig. 1.48 should allow bidirectional power transfer so that charging and discharging of batteries can be managed.

Electric vehicles (EVs) are increasingly connected to power grids through EV chargers. For demonstration, Fig. 1.49 shows a grid-tied architecture of EVs, where the rightmost dc–ac grid-tied converter serves as the grid interface of chargers. Despite that the majority of practical EV chargers are unidirectional for efficiency and lifetime improvement, the trend of using bidirectional EV chargers is obvious.

Besides being applied to renewable generation, ESSs, and EVs, we use power converters as grid interfaces for various power supplies and motor drivers, as mentioned. In addition, energy-efficient lighting requires grid-tied power converters. Importantly, power quality enhancement is mainly achieved through power electronic systems, such as uninterruptable power supplies (UPSs), dynamic voltage restorers (DVRs), and active power filters (APFs), as will be detailed. In medium-

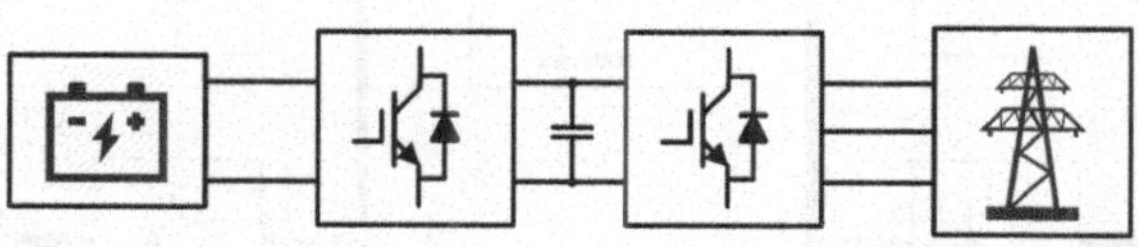

Fig. 1.48 Grid-tied architecture of battery storage systems

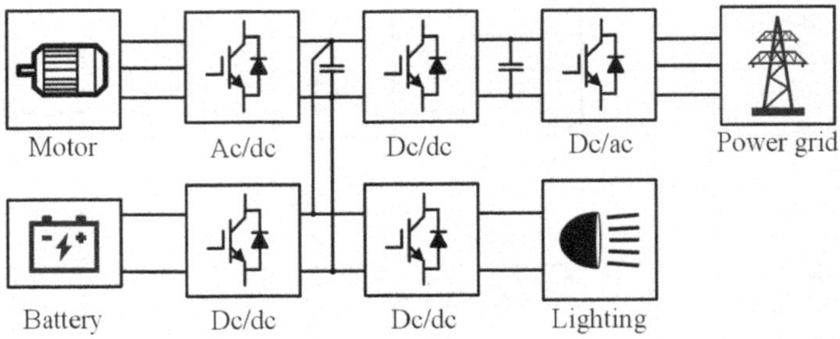

Fig. 1.49 Grid-tied architecture of electric vehicles

and high-voltage ac/dc transmission applications, grid-tied power converters act as major converter valves, which enable flexible power flow control.

The widespread use of power electronics as grid interfaces gradually shapes modern power systems into more-electronics power systems. Figure 1.50 demon-

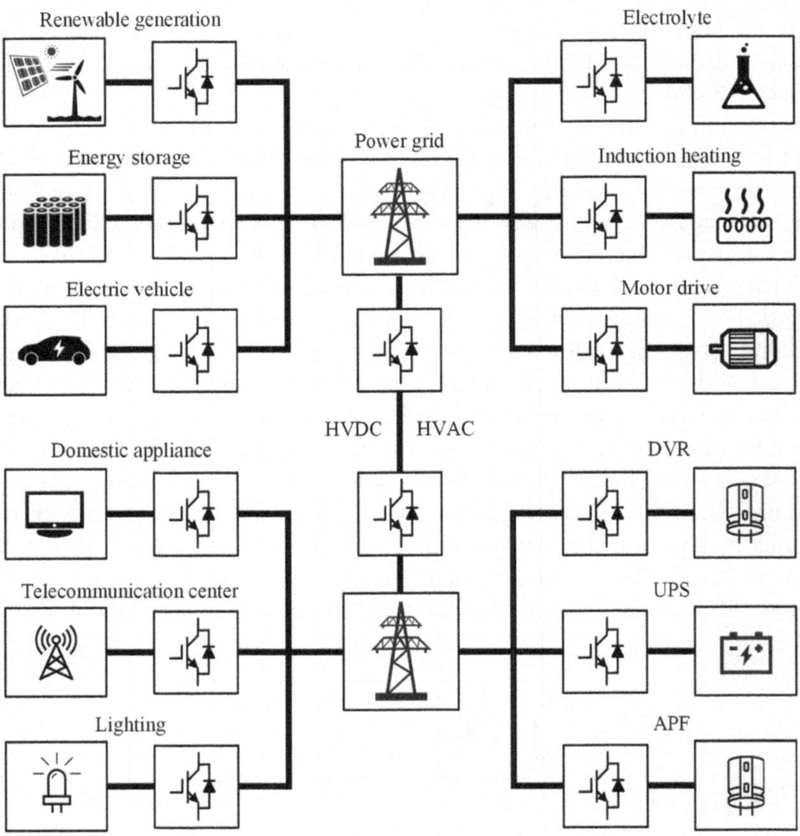

Fig. 1.50 Schematic of more-electronics power systems

strates a schematic of more-electronics power systems, which covers typical grid applications of power electronics. Notably, grid structures of more-electronics power systems vary enormously. However, more-electronics power systems share some common features.

1.4.2 *Key Features of More-Electronics Power Systems*

Power Electronics-Dominated Power Quality

Power quality issues refer to voltage and current deviations that cause malfunction or failure of customer equipment [45]. In more-electronics power systems, power quality is mainly determined by power electronics. Such a decisive role of power electronics is twofold.

First, power electronics should be blamed for the majority of power quality problems in more-electronics power systems, as will be detailed in the next chapter. Second, we can significantly boost power quality of more-electronics power systems through power quality conditioning equipment, which essentially refers to power electronic converters. We will elaborate on this in Chap. 3. In addition, it should be kept in mind that power converters as grid interfaces normally necessitate higher power quality, thereby pushing forward research and development of power quality standards and improvement approaches.

Wide-Spectrum Stability Challenges and Solutions

In addition to power quality challenges, the large-scale employment of power converters causes stability problems. In particular, stability problems of individual power converters (i.e., converter-level stability problems) may lead to system-level instability. Therefore, we must guarantee the stable operation of power converters.

Different from conventional power systems, more-electronics power systems may encounter instability over a wide spectrum due to high-frequency operations of power converters. Correspondingly, stability solutions target at different frequency bands in more-electronics power systems. Also, interactions among power converters, power converters and passive components, or power converters and power system equipment (e.g., SGs), should seriously be treated. We will discuss converter-level stability problems in Chap. 4.

Despite with increased stability challenges, more-electronics power systems feature various stability improvement strategies as well. This is mainly because of the fast and flexible control features of power converters. By exploiting such features, we may expect even better converter-level and system-level stability of more-electronics power systems.

Low Reserve and Inertia

Another important feature of more-electronics power systems is the retirement of SGs. Referring to Fig. 1.50, we note that most equipment and devices are connected to power grids through power converters instead of SGs. As a result, desirable features of SGs retire correspondingly.

As mentioned, system-level stability (of power systems) depends on the balance of active power and reactive power. In bulk power systems, SGs, together with their governors and excitors, undertake the responsibility of active and reactive power regulation. As SGs gradually retire, the maintenance of system-level stability becomes difficult. In addition, the levels of regulation reserve and inertia continue dropping, thus challenging the secure operation of more-electronics power systems. Once again, we tend to use power electronic equipment as a solution to these challenges.

High Penetration Level of Renewable Energy

Another feature that is closed related to the drop of regulation reserve and inertia refers to considerable power fluctuations caused by RESs. The stochastic and intermittent nature of RESs poses a threat on system-level stability of more-electronics power systems. We will handle this threat through power converters and energy storage, as will be discussed later.

We have introduced the concept and key features of more-electronics power systems. Regarding subtopics, all those related to power electronics and power systems are also subtopics in more-electronics power systems and hence will not be repeated here. As modern power systems evolve into more-electronics power systems, they will find applications not only as bulk power systems but also as small systems, such as microgrids.

In this chapter, we introduce the fundamentals of more-electronics power systems. First, we learn about power engineering. Second, we elaborate on power electronics, including its concept, key features, subtopics, and applications. Third, we detail power systems, particularly for modeling of power generation. Fourth, we derive more-electronics power systems and present their key features.

References

1. Newell WE (1974) Power electronics—emerging from limbo. IEEE Trans Ind Appl IA-10(1):7–11
2. Erickson RW, Maksimovic D (2001) Fundamentals of power electronics. Springer, New York, NY, USA
3. Fang J, Li H, Tang Y (2017) A magnetic integrated *LLCL* filter for grid-connected voltage-source converters. IEEE Trans Power Electron 32(3):1725–1730
4. Wang H, Blaabjerg F (2014) Reliability of capacitors for DC-link applications in power elecOtronic converters—an overview. IEEE Trans Ind Appl 50(5):3569–3578
5. Kossov OA (1968) Comparative analysis of chopper voltage regulators with LC filter. IEEE Trans Magnet 4(4):712–715
6. Ćuk S, Middlebrook RD (1977) A new optimum topology switching dc-to-dc converter. In: IEEE PESC, pp 160–179
7. Wester GW, Middlebrook RD (1972) Low-frequency characterization of switched dc-to-dc converters. In: IEEE PESC, pp 9–20
8. Middlebrook RD, Ćuk S (1976) A general unified approach to modeling switching converter power stages. In: IEEE PESC, pp 18–34

9. Brown AR, Middlebrook RD (1981) Sampled-data modeling of switching regulators. In: IEEE PESC, pp 349–369
10. Erikson RW, Ćuk S, Middlebrook RD (1982) Large-signal modeling and analysis of switching regulators. In: IEEE PESC, pp 240–250
11. Sanders SR, Noworolski JM, Liu XZ, Verghese GC (1991) Generalized averaging method for power conversion circuits. IEEE Trans Power Electron 6(2):251–259
12. Fang J, Yang X, Zhang L, Tang Y (2017) An optimal digital pulse-width-modulated dither technique to enhance the resolution of high frequency power converters. IEEE Trans Power Electron 32(9):7222–7232
13. Chen C (1999) Linear system theory and design. Oxford University Press, New York, NY, USA
14. Lee GT (2018) Abstract algebra—an introductory course. Springer, Cham, CH
15. Bengtsson G (1997) Output regulation and internal models—a frequency domain approach. Automatica 13(4):333–345
16. Zmood DN, Holmes DG (2003) Stationary frame current regulation of PWM inverters with zero steady-state error. IEEE Trans Power Electron 18:814–822
17. Dorf RC, Tallarida RJ (1993) Pocket book of electrical engineering formulas. CRC Press, Boca Raton, FL, USA
18. Fang J, Xiao G, Yang X, Tang Y (2017) Parameter design of a novel series-parallel-resonant *LCL* filter for single-phase half-bridge active power filters. IEEE Trans Power Electron 32(1):200–217
19. C2000 Delfino MCU F28379D LaunchPad™ development kit, Texas Instruments. https://www.ti.com/tool/LAUNCHXL-F28379D/
20. Tang Y, Blaabjerg F, Loh PC, Jin C, Wang P (2015) Decoupling of fluctuating power in single-phase systems through a symmetrical half-bridge circuit. IEEE Trans Power Electron 30:1855–1865
21. Fang J, Li X, Yang X, Tang Y (2017) An integrated trap–*LCL* filter with reduced current harmonics for grid-connected converters under weak grid conditions. IEEE Trans Power Electron 32(11):8446–8457
22. Sallen RP, Key EL (1955) A practical method of designing RC active filters. IRE Trans Circuit Theory 2(1):74–85
23. AMC 1303x small, high-precision, reinforced isolated delta-sigma modulators with internal clock, Texas Instruments. https://www.ti.com/product/AMC1303E0510?qgpn=amc1303
24. TMS320F2837xD Dual-Core Microcontrollers, Texas Instruments, Dallas, Texas, USA (2019)
25. Infineon EiceDRIVER™ gate driver ICs selection guide 2019, Infineon Technologies Austria AG 9500 Villach, Austria (2019)
26. Fang J, Tang Y, Li H, Li X (2018) A battery/ultracapacitor hybrid energy storage system for implementing the power management of virtual synchronous generators. IEEE Trans Power Electron 33(4):2820–2824
27. Global EV Outlook 2018, International Energy Agency. https://www.eesi.org/articles/view/comparing-u.s.-and-chinese-electric-vehicle-policies
28. Incentives for electric vehicles in Norway, Ecofys. https://www.euki.de/wp-content/uploads/2018/09/fact-sheet-incentives-for-electric-vehicles-no.pdf
29. Li H, Fang J, Chen S, Wang K, Tang Y (2018) Pulse density modulation for maximum efficiency point tracking of wireless power transfer systems. IEEE Trans Power Electron 33(6):5492–5501
30. Kundur P (1994) Power system stability and control. McGraw-Hill, New York, NY, USA
31. Tesla N (1888) A new system of alternate current motors and transformers. American Institute of Electrical Engineers, Piscataway, NJ, USA
32. Liu X, Wang P, Loh PC (2011) A hybrid AC/DC microgrid and its coordination control. IEEE Trans Smart Grid 2(2):278–286
33. Liu Y, You S, Liu Y (2017) Study of wind and PV frequency control in U.S. power grids—EI and TI case studies. IEEE Power Energy Technol J 4(3):65–73
34. AEMO (2017) Electricity rule change proposal, Aust. Energy Market Operator, Melbourne, VIC, Australia. https://www.aemo.com.au

35. EMA (2017) Enhancement to the spinning reserve requirements for the Singapore power system, Energy Market Authority, Singapore. https://www.ema.gov.sg
36. Nilsson JW, Riedel SA (2015) Electric circuits, 10th edn. Prentice Hall, New Jersey, NJ, USA
37. Fang J, Lin P, Li H, Yang Y, Tang Y (2019) An improved virtual inertia control for three-phase voltage source converters connected to a weak grid. IEEE Trans Power Electron 34(9):8660–8670
38. Fang J, Li H, Tang Y, Blaabjerg F (2019) On the inertia of future more-electronics power systems. IEEE J Emerg Sel Topics Power Electron 7(4):2130–2146
39. Li Y, Vilathgamuwa DM, Loh PC (2004) Design, analysis, and real-time testing of a controller for multibus microgrid system. IEEE Trans Power Electron 19(5):1195–1204
40. Guerrero JM, Vasquez JC, Matas J, de Vicuna LG, Castilla M (2011) Hierarchical control of droop-controlled AC and DC microgrids—a general approach toward standardization. IEEE Trans Ind Electron 58(1):158–172
41. Loh PC, Li D, Chai YK, Blaabjerg F (2013) Autonomous operation of hybrid microgrid with ac and dc subgrids. IEEE Trans Power Electron 28(5):2214–2223
42. Amin SM, Wollenberg BF (2008) Toward a smart grid. IEEE Mag Power Energy 8(5):114–122
43. Fang J, Li X, Tang Y, Li H (2017) Improvement of frequency stability in power electronics-based power systems. In: Proceedings of the ACEPT, pp 1–6, Singapore, 24–26 Oct. 2017
44. REN21, Paris, France, Renewables 2017 global status report (2017) https://www.ren21.net
45. Dugan RC, McGranaghan MF, Santoso S, Beaty HW (2003) Electrical power systems quality. McGraw-Hill, 2 Penn Plaza, New York, NY, USA

Chapter 2
Power Quality Problems and Standards

2.1 Voltage Quality Problems

As introduced, power quality problems refer to the problems that cause malfunction or failure of customer equipment. In general, power quality problems are divided into voltage-quality problems and current-quality problems. Normally, power system operators should be responsible for voltage-quality problems, as voltage problems are related to the voltages supplied by grids. In contrast, current quality problems are generally caused by loads. However, voltage-quality problems and current-quality problems do affect each other. This section focuses on voltage-quality problems, covering voltage amplitude deviations, harmonics, and imbalances.

2.1.1 Voltage Amplitude Deviations

Remember that ac quantities feature three important properties—amplitude, frequency, and phase angle. Correspondingly, power system operators should maintain grid voltage amplitude, frequency, and phase angle within their allowable ranges. Specifically, voltage amplitude deviations refer to negative/positive voltage amplitude changes. The fundamental frequency of voltages is closely related to system-level stability and hence will be discussed later. In addition, grid voltages may contain high-frequency harmonics, leading to distortions of voltage waveforms, which is covered in this chapter. Additionally, incorrect phase relationships among three-phase voltages cause voltage imbalances. Next, we will discuss voltage quality problems, starting with voltage amplitude problems in this subsection.

In the previous chapter, we have shown rated voltage amplitudes of several power systems in Table 1.5. Specifically, rated voltages range from 100 to 230 V (in terms of phase-to-ground root-mean-square voltages). Normally, grid voltages should be

J. Fang, *More-Electronics Power Systems: Power Quality and Stability*,
Power Systems, https://doi.org/10.1007/978-981-15-8590-6_2

maintained within $\pm 10\%$. Otherwise, customer equipment may malfunction. Thus, special care should be exerted to deviations of grid voltage amplitudes.

On the one hand, grid voltages may be generated with errors. For instance, if excitation currents are improper, we know from the stead-state model of SGs that their stator voltage amplitudes can deviate, leading to grid-voltage amplitude errors. In more-electronics power systems, grid-tied power converters as generators may also cause voltage amplitude deviations, e.g. due to improper control.

On the other hand, electric loads may cause grid voltage deviations. In this case, generators produce standard voltages. However, load currents flowing along transmission lines introduce voltage drops, thereby leading to grid voltage deviations seen from the load side. In fact, this problem is the greatest contributor that leads to the failure of dc power systems in the early years. In this sense, the majority of voltage deviations are induced by loads instead of sources, which is proven to be true [1]. Therefore, power grids set strict standards for loads to make sure that they do not produce excessively undesirable currents.

In terms of classifications, voltage amplitude deviations are further divided into several sub-problems based on amplitudes and time durations. However, specific definitions may differ according to various standards.

Overvoltage

First, we consider long-term voltage deviations, where voltage deviations span longer than one minute. We define an *overvoltage* as the voltage with its amplitude deviation exceeding 10% in the positive direction for more than one minute. As a result, over voltages damage customer equipment in the long term.

In bulk power systems, long-term voltage amplitudes largely depend on the balance of reactive power. Therefore, long-term voltage deviations, e.g., over voltages, can be caused by improper reactive power control. In particular, excessive reactive generation leads to over voltages.

Undervoltage

Corresponding to the overvoltage, an *undervoltage* is defined as the negative voltage amplitude deviation that exceeds 10% and lasts for more than one minute. By definition, system blackouts (i.e., with zero voltages) can be classified into undervoltage events. Generally, under voltages prevent customer equipment from normal operation instead of damaging them.

Voltage Swell

A *voltage swell* refers to a positive voltage deviation that exceeds 10%, and its time duration ranges from a half of the fundamental period to one minute. Notably, voltage swells and over voltages are separated in terms of time scales. In comparison, voltage swells feature faster dynamics than over voltages. Moreover, voltage swells are often with greater voltage deviations.

Voltage Sag

Dual to the voltage swell, a *voltage sag* (or a *voltage dip*) refers to a voltage drop that exceeds 10% and spans from 0.5 fundamental period to one minute. In most scenarios, voltage sags refer to voltages with amplitudes greater than 0.1 rated values. Such a requirement helps to differentiate voltage sags from blackouts.

In practice, the majority of voltage problems are voltage sags. Therefore, heightened attention should be paid to voltage sags. In more-electronics power systems, grid-tied power converters are increasingly required to work properly under voltage sags. The ability of power converters to ride through voltage sags is the key aspect of fault-ride through.

Voltage Flicker

A *voltage flicker* refers to a random yet small voltage deviation (normally within ±10%) that introduces visible changes of light levels. Generally, voltage flickers are not as serious as the previously mentioned voltage quality problems. However, they might bring troubles to sensitive loads and human being. Therefore, it is desirable to remove voltage flickers.

Voltage Spike

A *voltage spike* (also known as a *voltage impulse* or *voltage surge*) means a very fast voltage change. A sudden change of loads (e.g., turning off motor loads) may lead to voltage spikes. In most cases, voltage spikes are not serious voltage quality problems. However, they may bring negative effects to sensitive loads.

Table 2.1 summarizes voltage quality problems related to voltage amplitudes, where the problems are differentiated by voltage amplitude changes and time durations. It should be remembered that the relevant amplitude or time data may differ according to various standards without loss of generality. In general, voltage deviations exceeding ±10% are not allowable in most power systems.

For demonstration, Fig. 2.1 presents the experimental waveforms of a grid voltage sag, where the voltage amplitude reduces about 10% and lasts for 3.2 s. During the voltage sag, grid-tied power converters should not be disconnected from power grids for protection and should preferably contribute more reactive power for grid support. Solutions to voltage quality problems, e.g., compensation of voltage sags, will be detailed in the next chapter.

Table 2.1 Voltage quality problems related to voltage amplitudes

Voltage quality problems	Amplitude changes	Time durations
Voltage sags	$< -10\%$	10 ms – 1 min
Voltage swells	$> +10\%$	10 ms – 1 min
Under voltages	$< -10\%$	>1 min
Over voltages	$> +10\%$	>1 min
Voltage flickers	Within ±10%	≥10 ms
Voltage spikes	$> +10\%$	<10 ms

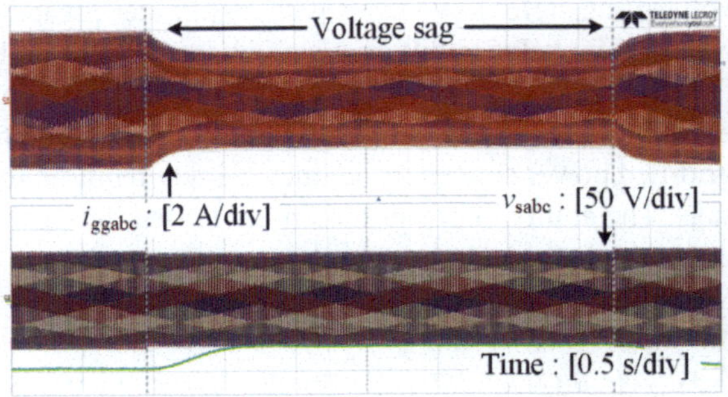

Fig. 2.1 Experimental waveforms of a voltage sag

2.1.2 Voltage Harmonics

As known, fundamental voltages are of the line frequency (i.e., 50/60 Hz for bulk power systems). *Harmonics* are components whose frequencies are multiples of the line frequency. With a fundamental frequency of 50 Hz, harmonics are of frequencies 50*n* Hz (where *n* is an integer), e.g., 100, 150, 200 Hz, etc.

According to Fourier series, any practical periodical signal can be expanded into a linear combination of multiple sinusoidal and dc components. To be specific, sinusoidal components include fundamental component and harmonics. The relative amplitudes and phases of harmonics determine the shape of periodical signals. In other words, harmonics tell how "sinusoidal" periodical signals are. To evaluate the sinusoidal degree of signals, the terminology *total harmonics distortion* (THD) is defined as the ratio of the root-sum-square of harmonics and the fundamental component. For example, the voltage THD is calculated to be

$$\mathrm{THD_v} = \frac{\sqrt{V_{\mathrm{s\_rms}}^2 - V_{\mathrm{o\_rms}}^2}}{V_{\mathrm{o\_rms}}}, \tag{2.1}$$

where $V_{\mathrm{o\_rms}}$ stands for the root-mean-square (rms) value of the fundamental voltage component, while $V_{\mathrm{s\_rms}}$ represents the rms of the voltage signal. Clearly, the numerator of voltage THDs quantifies harmonic components. Therefore, a large THD translates into more harmonic distortions. To maintain sinusoidal waveforms, grid voltage THDs should be maintained sufficiently low (e.g., less than 5%).

Figure 2.2 illustrates the experimental waveforms of power converters under a distorted grid environment, where grid voltages contain 10% 5th and 4% 7th harmonics. Normally, odd-order harmonics are worse than even-order harmonics. In three-phase systems, $6n \pm 1$ (where n is an integer) harmonics dominate.

Notably, power converters are significant harmonic sources [2]. As an example, diode rectifiers generate harmonic currents, which translate into voltage harmonics after propagation along power lines. Figure 2.3 shows the schematic of diode rectifiers, where the left-hand-side circuit refers to a single-phase H-bridge rectifier, while the right-hand-side rectifier is widely used in three-phase systems.

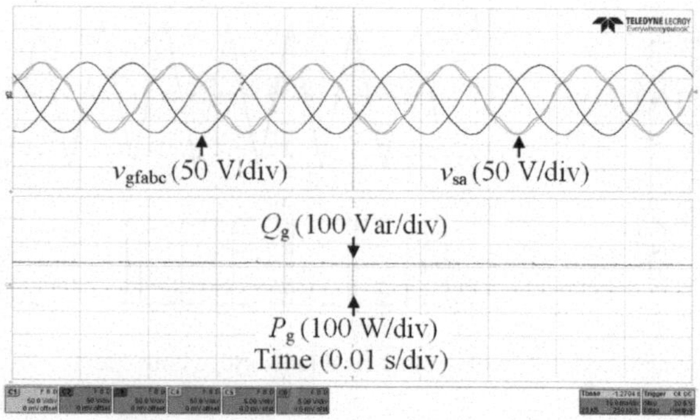

Fig. 2.2 Experimental waveforms of power converters under distorted grid voltages

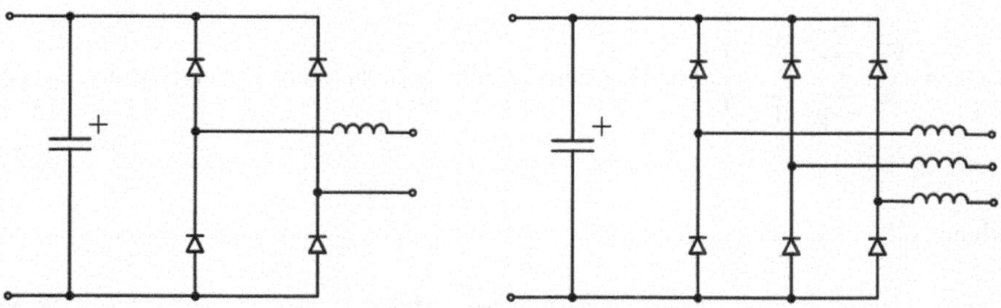

Fig. 2.3 Schematic of diode rectifiers

Components with frequencies located between harmonic frequencies (e.g., 125 Hz) are defined as inter-harmonics. Generally, inter-harmonics are much weaker than harmonics in terms of amplitudes. Therefore, we mainly focus on harmonics instead of inter-harmonics.

Generally, harmonics and inter-harmonics affect more-electronics power systems negatively. Without proper treatment, they may cause overheating, efficiency drop, noise, equipment malfunction and failure, and even fire. Therefore, we must attenuate harmonics and inter-harmonics to comply with grid codes.

Extra care should be exerted to switching harmonics introduced by power converters in more-electronics power systems. Noticeably, switching behaviors of power converters generate the switching harmonics that have a tight bearing on the switching frequency. Dependent on modulation schemes and topologies, switching harmonics can appear at the switching frequency and/or its multiples. Taking symmetrical half-bridge converters as an example, we present their schematic and natural sampling modulation scheme in Fig. 2.4, where S_{iu} and S_{id} refer to the upper and lower switches of converters, respectively. m_i represents the reference sinusoidal waveform. After compared with a triangular waveform, m_i produces the driving signal g_{su} that determines the on–off actions of S_{iu}. Note that S_{id} operates complimentarily with S_{iu}.

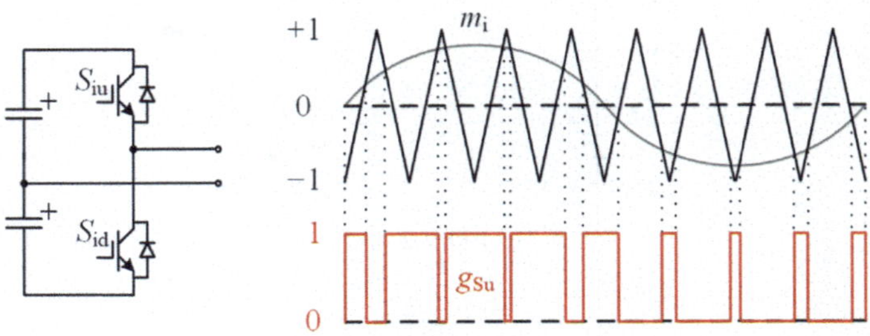

Fig. 2.4 Schematic and modulation of symmetrical half-bridge converters

Given that the symmetrical-regular-sampled PWM (which samples m_i once each switching period) is involved, the converter voltage signal is expressed as [3]

$$v_s(t) = \frac{V_{dc}}{\pi} \sum_{\substack{m=0 \\ m>0}}^{\infty} \sum_{\substack{n=1 \\ n=-\infty}}^{\infty} \frac{1}{q} J_n(qM\frac{\pi}{2}) \sin\left[(q+n)\frac{\pi}{2}\right] \cos[m(\omega_c t + \theta_c) + n(\omega_o t + \theta_o)], \tag{2.2}$$

where

$$q = m + \frac{n\omega_o}{\omega_c} \text{ and} \tag{2.3}$$

$$J_n(x) = \frac{1}{2\pi} \int_0^{2\pi} \cos(nt - x \sin t)dt. \tag{2.4}$$

In (2.2)–(2.4), $v_s(t)$ refers to the converter output voltage, V_{dc} stands for the upper or lower capacitor voltage, and m and n designate the orders of the modulation wave and carrier wave, respectively. M is defined as the modulation ratio. In addition, ω_c and ω_o represent the angular frequencies of the carrier wave and fundamental component, respectively. Similarly, θ_c and θ_o nominate the phase angles of the carrier wave and fundamental component, respectively. The function $J_n()$ in (2.4) is known as the Bessel function [3]. It should be noted that the absolute value of the coefficients before the cosine term in (2.2) is essentially the amplitude of switching harmonics.

As an example, we select $M = 0.9$, $\omega_c/\omega_o = 21$, and $\theta_c = \theta_o = 0°$ and plot the voltage spectrum of symmetrical half-bridge converters in Fig. 2.5. We find that voltage harmonics mainly appear around the switching frequency (i.e., the frequency of the 21st harmonic) and its multiples. Moreover, low-frequency harmonics are negligible. Besides, the magnitudes of odd-order (or even-order) sidebands around odd-order (or even-order) carriers are insignificant [3].

As switching harmonics are of high frequencies beyond the bandwidth of controllers, they are typically attenuated by passive filters of power converters. We

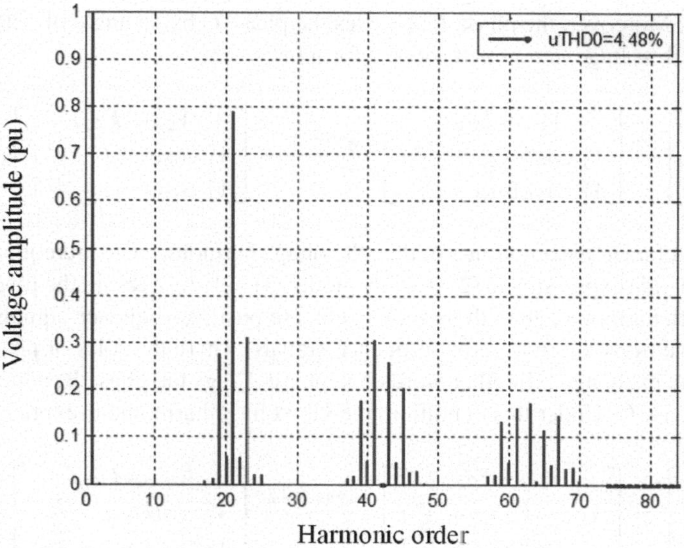

Fig. 2.5 Voltage spectrum of symmetrical half-bridge converters

will further discuss the harmonic suppression effect of passive filters in the next chapter.

2.1.3 Voltage Imbalances

In balanced three-phase systems, three-phase grid voltages are expected to be evenly shifted with $\pm 120°$ phase angle differences and identical voltage amplitudes. Alternatively, *voltage imbalances* are characterized by either amplitude inconsistency or undesirable phase angle differences. Voltage imbalances may cause malfunction or even failure of customer equipment.

Three-phase voltages are coupled to each other. As such, a voltage deviation in one phase may lead to voltage imbalances. Voltage imbalances are often caused by unsymmetrical loads, mismatched power lines, and/or system faults. In this sense, power system operators prefer to distribute single-phase loads as evenly as possible. To quantify voltage imbalances, we will expand voltage signals into three signals with different sequences—positive sequence, negative sequence, and zero sequence.

2.1.3.1 Sequences of Harmonics

To learn sequences, let us take a single balanced kth voltage harmonic as an example. First of all, we look at positive-sequence signals, where the phase A leads the phase

B by 120°. Moreover, the phase B precedes the phase C by an angle of 120°. In this case, the kth voltage harmonic takes the form of

$$\begin{bmatrix} v_{\text{ska}} \\ v_{\text{skb}} \\ v_{\text{skc}} \end{bmatrix} = \begin{bmatrix} V_{\text{k}} \cos k\omega_{\text{o}}t \\ V_{\text{k}} \cos k(\omega_{\text{o}}t - \frac{2\pi}{3}) \\ V_{\text{k}} \cos k(\omega_{\text{o}}t + \frac{2\pi}{3}) \end{bmatrix} \xrightarrow{\text{mod}(k,3)=1} \begin{bmatrix} V_{\text{k}} \cos k\omega_{\text{o}}t \\ V_{\text{k}} \cos(k\omega_{\text{o}}t - \frac{2\pi}{3}) \\ V_{\text{k}} \cos(k\omega_{\text{o}}t + \frac{2\pi}{3}) \end{bmatrix}, \quad (2.5)$$

where V_{k} refers to the amplitude of the kth voltage harmonic. The prerequisite of the right-hand-side expression in (2.5) is that mod(k, 3) = 1. As a result, the fundamental voltage, 4th harmonic, and 7th harmonic, etc. are positive-sequence signals.

Next, we move on to negative-sequence signals, where the roles of phase A and phase C are exchanged. In other words, the phase A lags the phase B, which further lags the phase C. Under this condition, the kth voltage harmonic is expressed as

$$\begin{bmatrix} v_{\text{ska}} \\ v_{\text{skb}} \\ v_{\text{skc}} \end{bmatrix} = \begin{bmatrix} V_{\text{k}} \cos k\omega_{\text{o}}t \\ V_{\text{k}} \cos k(\omega_{\text{o}}t - \frac{2\pi}{3}) \\ V_{\text{k}} \cos k(\omega_{\text{o}}t + \frac{2\pi}{3}) \end{bmatrix} \xrightarrow{\text{mod}(k,3)=2} \begin{bmatrix} V_{\text{k}} \cos k\omega_{\text{o}}t \\ V_{\text{k}} \cos(k\omega_{\text{o}}t + \frac{2\pi}{3}) \\ V_{\text{k}} \cos(k\omega_{\text{o}}t - \frac{2\pi}{3}) \end{bmatrix}, \quad (2.6)$$

where mod(k, 3) = 2 should be satisfied. Therefore, 2nd harmonic, 5th harmonic, and 8th harmonic, etc. belong to negative-sequence signals.

Finally, zero-sequence signals are of identical phases. Correspondingly, the kth voltage harmonic is described by

$$\begin{bmatrix} v_{\text{ska}} \\ v_{\text{skb}} \\ v_{\text{skc}} \end{bmatrix} = \begin{bmatrix} V_{\text{k}} \cos k\omega_{\text{o}}t \\ V_{\text{k}} \cos k(\omega_{\text{o}}t - \frac{2\pi}{3}) \\ V_{\text{k}} \cos k(\omega_{\text{o}}t + \frac{2\pi}{3}) \end{bmatrix} \xrightarrow{\text{mod}(k,3)=0} \begin{bmatrix} V_{\text{k}} \cos k\omega_{\text{o}}t \\ V_{\text{k}} \cos k\omega_{\text{o}}t \\ V_{\text{k}} \cos k\omega_{\text{o}}t \end{bmatrix}, \quad (2.7)$$

where three-phase voltages overlap. Zero-sequence signals include dc component, 3rd harmonic, 6th harmonic, and so on. Notably, harmonics are naturally classified into three sequences. For instance, it is expected that the balanced three-phase 2nd harmonic is of a negative sequence. In other words, if the 2nd harmonic contains positive-sequence signals, it is claimed that voltage imbalances exist. Table 2.2 summarizes the sequences of normal (or balanced) harmonics.

Table 2.2 Sequences of normal voltage harmonics

Harmonic orders	Sequences	Examples
$3n + 1$	Positive	Fundamental voltage, 4th harmonic, 7th harmonic, ...
$3n - 1$	Negative	2nd harmonic, 5th harmonic, 8th harmonic, ...
$3n$	Zero	3rd harmonic, 6th harmonic, 9th harmonic, ...

2.1.3.2 Sequence Decomposition of Harmonics

For imbalanced three-phase harmonics, each of them can be decomposed into a sum of positive-sequence, negative-sequence, and zero-sequence harmonic components. To illustrate this, we first express the kth voltage harmonic in a phasor form as

$$\begin{bmatrix} v_{\mathrm{ska}} \\ v_{\mathrm{skb}} \\ v_{\mathrm{skc}} \end{bmatrix} = \begin{bmatrix} V_{\mathrm{a}} \cos k(\omega_{\mathrm{o}} t + \theta_{\mathrm{a}}) \\ V_{\mathrm{b}} \cos k(\omega_{\mathrm{o}} t + \theta_{\mathrm{b}}) \\ V_{\mathrm{c}} \cos k(\omega_{\mathrm{o}} t + \theta_{\mathrm{c}}) \end{bmatrix} \Leftrightarrow \begin{bmatrix} V_{\mathrm{a}} e^{j\theta_{\mathrm{a}}} \\ V_{\mathrm{b}} e^{j\theta_{\mathrm{b}}} \\ V_{\mathrm{c}} e^{j\theta_{\mathrm{c}}} \end{bmatrix}, \tag{2.8}$$

where V_{a}, V_{b}, and V_{c} represent the three-phase voltage amplitudes of harmonics, respectively. Accordingly, θ_{a}, θ_{b}, and θ_{c} denote the three-phase angles of harmonics, respectively.

Next, we assume that the kth voltage harmonic can be decomposed into

$$\begin{bmatrix} V_{\mathrm{a}} e^{j\theta_{\mathrm{a}}} \\ V_{\mathrm{b}} e^{j\theta_{\mathrm{b}}} \\ V_{\mathrm{c}} e^{j\theta_{\mathrm{c}}} \end{bmatrix} = \begin{bmatrix} V_{\mathrm{a1}} e^{j\theta_1} \\ V_{\mathrm{a1}} e^{j(\theta_1 - \frac{2}{3}\pi)} \\ V_{\mathrm{a1}} e^{j(\theta_1 + \frac{2}{3}\pi)} \end{bmatrix} + \begin{bmatrix} V_{\mathrm{a2}} e^{j\theta_2} \\ V_{\mathrm{a2}} e^{j(\theta_2 + \frac{2}{3}\pi)} \\ V_{\mathrm{a2}} e^{j(\theta_2 - \frac{2}{3}\pi)} \end{bmatrix} + \begin{bmatrix} V_{\mathrm{a0}} e^{j\theta_0} \\ V_{\mathrm{a0}} e^{j\theta_0} \\ V_{\mathrm{a0}} e^{j\theta_0} \end{bmatrix}, \tag{2.9}$$

where V_{a1}, V_{a2}, and V_{a0} refer to the voltage amplitudes of positive-sequence, negative-sequence, and zero-sequence components, respectively. θ_1, θ_2, and θ_0 are the relevant phase angles.

To proof (2.9), we define a phasor rotation operator α as

$$\alpha \equiv e^{j\frac{2}{3}\pi}, \tag{2.10}$$

which implies that

$$\alpha^3 = 1, \alpha = \frac{1}{\alpha^2}, \alpha^2 = \alpha^{-1}. \tag{2.11}$$

Using (2.10) and (2.11), we reorganize the right-hand-side terms of (2.9) as

$$\begin{bmatrix} V_{\mathrm{a1}} e^{j\theta_1} \\ V_{\mathrm{a1}} e^{j(\theta_1 - \frac{2}{3}\pi)} \\ V_{\mathrm{a1}} e^{j(\theta_1 + \frac{2}{3}\pi)} \end{bmatrix} = V_{\mathrm{a1}} e^{j\theta_1} \begin{bmatrix} 1 \\ \alpha^2 \\ \alpha \end{bmatrix}, \tag{2.12}$$

$$\begin{bmatrix} V_{\mathrm{a2}} e^{j\theta_2} \\ V_{\mathrm{a2}} e^{j(\theta_2 + \frac{2}{3}\pi)} \\ V_{\mathrm{a2}} e^{j(\theta_2 - \frac{2}{3}\pi)} \end{bmatrix} = V_{\mathrm{a2}} e^{j\theta_2} \begin{bmatrix} 1 \\ \alpha \\ \alpha^2 \end{bmatrix}, \text{ and} \tag{2.13}$$

$$\begin{bmatrix} V_{\mathrm{a0}} e^{j\theta_0} \\ V_{\mathrm{a0}} e^{j\theta_0} \\ V_{\mathrm{a0}} e^{j\theta_0} \end{bmatrix} = V_{\mathrm{a0}} e^{j\theta_0} \begin{bmatrix} 1 \\ 1 \\ 1 \end{bmatrix}. \tag{2.14}$$

Substitution of (2.12)–(2.14) into (2.9), the latter is reorganized as

$$\begin{bmatrix} V_a e^{j\theta_a} \\ V_b e^{j\theta_b} \\ V_c e^{j\theta_c} \end{bmatrix} = \begin{bmatrix} 1 & 1 & 1 \\ \alpha^2 & \alpha & 1 \\ \alpha & \alpha^2 & 1 \end{bmatrix} \begin{bmatrix} V_{a1} e^{j\theta_1} \\ V_{a2} e^{j\theta_2} \\ V_{a0} e^{j\theta_0} \end{bmatrix}. \tag{2.15}$$

Finally, we achieve sequence decomposition by multiplying the inverse matrix of (2.15) on both sides. Besides, we change the left-hand-side and right-hand-side expressions, leading to

$$\begin{bmatrix} V_{a1} \cos k(\omega_o t + \theta_1) \\ V_{a2} \cos k(\omega_o t + \theta_2) \\ V_{a0} \cos k(\omega_o t + \theta_0) \end{bmatrix} \Leftrightarrow \begin{bmatrix} V_{a1} e^{j\theta_1} \\ V_{a2} e^{j\theta_2} \\ V_{a0} e^{j\theta_0} \end{bmatrix} = \frac{1}{3} \begin{bmatrix} 1 & \alpha & \alpha^2 \\ 1 & \alpha^2 & \alpha \\ 1 & 1 & 1 \end{bmatrix} \begin{bmatrix} V_a e^{j\theta_a} \\ V_b e^{j\theta_b} \\ V_c e^{j\theta_c} \end{bmatrix}, \tag{2.16}$$

which tells how to derive balanced three-phase harmonic components from imbalanced signals. In particular, the sum of original three-phase signals with reverse phase shifts gives rise to the positive-sequence component. In contrast, the sum of original three-phase signals with normal phase shifts leads to the negative-sequence component. The direct sum of three-phase signals yields the zero-sequence component. It should be emphasized that the frequency of phasors are identical to those of harmonics.

2.1.3.3 Voltage Imbalances

As mentioned, each periodical voltage can be expanded into a series of dc component and sinusoidal signals, including the fundamental component and voltage harmonics. Further, the fundamental component or each harmonic can be decomposed into positive-sequence, negative-sequence, and zero-sequence harmonic components. As a result, the periodical voltage is expressed as the sum of positive-sequence, negative-sequence, and zero-sequence harmonic components.

On the basis of sequence decomposition, the *voltage imbalance* (or *voltage unbalance*) is defined as the ratio of the negative-sequence and/or zero-sequence components to the positive-sequence component. Note that voltage imbalances only exist in poly-phase systems. In practice, we evaluate voltage imbalances through the following experience formula:

$$\frac{\max(V_a, V_b, V_c) - \text{ave}(V_a, V_b, V_c)}{\text{ave}(V_a, V_b, V_c)}, \tag{2.17}$$

where the functions max() and ave() calculate the maximum and average values of their input elements, respectively. Generally, voltage imbalances are limited to be 1–3% by utilities.

2.1.3.4 Sequences and Frame Transformations

At this stage, it is relevant to investigate the influences of *abc/dq*0 and *dq*0/*abc* frame transformations on harmonic sequence components. We have already introduced frame transformations in the previous chapter. As mentioned, the *abc/dq*0 transformation reshapes three-phase-balanced grid voltages into constants in the *d*-axis. Furthermore, this part will explore the cases where three-phase voltages are imbalanced and/or contaminated with harmonics. In general, the orders and sequences of harmonics may change after transformations, thereby complicating systems analysis and design. However, by exploitation of the regular pattern of sequence and order changes, we can resolve problems and design systems in a simple and smart fashion.

Through the standard *abc/dq*0 transformation, a positive-sequence harmonic [see (2.5)] becomes

$$\frac{2}{3}\begin{bmatrix} \cos\omega_o t & \cos(\omega_o t-\frac{2\pi}{3}) & \cos(\omega_o t+\frac{2\pi}{3}) \\ -\sin\omega_o t & -\sin(\omega_o t-\frac{2\pi}{3}) & -\sin(\omega_o t+\frac{2\pi}{3}) \\ \frac{1}{2} & \frac{1}{2} & \frac{1}{2} \end{bmatrix}\begin{bmatrix} V_k\cos k\omega_o t \\ V_k\cos(k\omega_o t-\frac{2\pi}{3}) \\ V_k\cos(k\omega_o t+\frac{2\pi}{3}) \end{bmatrix} = \begin{bmatrix} V_k\cos(k-1)\omega_o t \\ V_k\sin(k-1)\omega_o t \\ 0 \end{bmatrix}, \tag{2.18}$$

which indicates that the order of harmonics reduces by one after transformation. This explains why fundamental voltages are changed into constants in the *d*-axis. Substitution of $k=1$ into (2.18), the *q*-axis component is found to be zero. In addition, the other positive-sequence harmonics are changed into zero-sequence order signals. For example, the 4th and 7th harmonics are transformed into the 3rd and 6th signals, respectively.

Correspondingly, a negative-sequence harmonic [see (2.6)] becomes

$$\frac{2}{3}\begin{bmatrix} \cos\omega_o t & \cos(\omega_o t-\frac{2\pi}{3}) & \cos(\omega_o t+\frac{2\pi}{3}) \\ -\sin\omega_o t & -\sin(\omega_o t-\frac{2\pi}{3}) & -\sin(\omega_o t+\frac{2\pi}{3}) \\ \frac{1}{2} & \frac{1}{2} & \frac{1}{2} \end{bmatrix}\begin{bmatrix} V_k\cos k\omega_o t \\ V_k\cos(k\omega_o t+\frac{2\pi}{3}) \\ V_k\cos(k\omega_o t-\frac{2\pi}{3}) \end{bmatrix} = \begin{bmatrix} V_k\cos(k+1)\omega_o t \\ -V_k\sin(k+1)\omega_o t \\ 0 \end{bmatrix}, \tag{2.19}$$

which demonstrates that the transformation increases the order of voltage harmonics by one. As noticed, the resultant *d*-axis and *q*-axis components are with identical amplitudes and frequency. In comparison, the *q*-axis component leads the *d*-axis component by 90°. As the order of harmonics increases by one, negative-sequence harmonics are transformed into zero-sequence order signals. For instance, the 2nd and 5th harmonics are transformed into the 3rd and 6th harmonics, respectively.

Table 2.3 *Abc/dq*0 transformations of normal voltage harmonics

Original harmonic orders	Sequences	Harmonic orders after transformation
$3n+1$	Positive	$3n$
$3n-1$	Negative	$3n$
$3n$	Zero	$3n$

Remember that 3rd and 6th harmonics may also be transformed from positive-sequence harmonics. Therefore, positive- and negative-sequence harmonics can be transformed into the same zero-sequence order signal. In other words, we cannot differentiate positive-sequence harmonics from negative-sequence harmonics (of different orders) after *abc*/*dq*0 transformations.

Further, we look at zero-sequence harmonics, which changes into

$$\frac{2}{3}\begin{bmatrix} \cos\omega_o t & \cos(\omega_o t-\frac{2\pi}{3}) & \cos(\omega_o t+\frac{2\pi}{3}) \\ -\sin\omega_o t & -\sin(\omega_o t-\frac{2\pi}{3}) & -\sin(\omega_o t+\frac{2\pi}{3}) \\ \frac{1}{2} & \frac{1}{2} & \frac{1}{2} \end{bmatrix}\begin{bmatrix} V_k\cos k\omega_o t \\ V_k\cos k\omega_o t \\ V_k\cos k\omega_o t \end{bmatrix} = \begin{bmatrix} 0 \\ 0 \\ V_k\cos k\omega_o t \end{bmatrix}, \tag{2.20}$$

where the zero-sequence harmonic is directly mapped into the same zero-sequence order component. In conclusion, all the normal (or balanced) harmonics are transformed into zero-sequence order harmonics, as summarized in Table 2.3.

However, it should be kept in mind that imbalanced harmonics may contain all three different sequence components. For example, the 3rd harmonic may comprise not only a zero-sequence component but also a positive-sequence component and a negative-sequence component. Correspondingly, the zero-sequence component remains unchanged after transformation. However, the positive-sequence component becomes 2nd harmonic components in both the *d*- and *q*-axes after transformation, where the *q*-axis component lags 90°. Alternatively, the negative-sequence component is changed into 4th harmonics in both the *d*- and *q*-axes after transformation, where the *q*-axis component leads 90°.

Next, we proceed to inverse transformations. We derive from (2.18) that

$$\begin{bmatrix} \cos\omega_O t & -\sin\omega_O t & 1 \\ \cos(\omega_O t-\frac{2\pi}{3}) & -\sin(\omega_O t-\frac{2\pi}{3}) & 1 \\ \cos(\omega_O t+\frac{2\pi}{3}) & -\sin(\omega_O t+\frac{2\pi}{3}) & 1 \end{bmatrix}\begin{bmatrix} V_k\cos k\omega_O t \\ V_k\sin k\omega_O t \\ 0 \end{bmatrix} = \begin{bmatrix} V_k\cos(k+1)\omega_O t \\ V_k\cos\left[(k+1)\omega_O t-\frac{2\pi}{3}\right] \\ V_k\cos\left[(k+1)\omega_O t+\frac{2\pi}{3}\right] \end{bmatrix}, \tag{2.21}$$

Table 2.4 *Dq0/abc* transformations of normal synchronous signals

Original harmonic orders	Features	Harmonic orders after transformation
$3n$	Q-axis component lags 90°	$3n + 1$
$3n$	Q-axis component leads 90°	$3n - 1$
$3n$	Only zero-axis component	$3n$

which increases the order of *dq* components by one and gives positive-sequence voltage signals. Likewise, we multiply the inverse transformation matrix on both sides of (2.19) and change the relevant indices, yielding

$$\begin{bmatrix} \cos\omega_0 t & -\sin\omega_0 t & 1 \\ \cos(\omega_0 t - \frac{2\pi}{3}) & -\sin(\omega_0 t - \frac{2\pi}{3}) & 1 \\ \cos(\omega_0 t + \frac{2\pi}{3}) & -\sin(\omega_0 t + \frac{2\pi}{3}) & 1 \end{bmatrix} \begin{bmatrix} V_k \cos k\omega_0 t \\ -V_k \sin k\omega_0 t \\ 0 \end{bmatrix} = \begin{bmatrix} V_k \cos(k-1)\omega_0 t \\ V_k \cos\left[(k-1)\omega_0 t + \frac{2\pi}{3}\right] \\ V_k \cos\left[(k-1)\omega_0 t - \frac{2\pi}{3}\right] \end{bmatrix}, \tag{2.22}$$

which reduces the order of *dq* components by one and changes them into negative-sequence voltage signals. Similarly, we derive the inverse transformation that gives zero-sequence signals as

$$\begin{bmatrix} \cos\omega_0 t & -\sin\omega_0 t & 1 \\ \cos(\omega_0 t - \frac{2\pi}{3}) & -\sin(\omega_0 t - \frac{2\pi}{3}) & 1 \\ \cos(\omega_0 t + \frac{2\pi}{3}) & -\sin(\omega_0 t + \frac{2\pi}{3}) & 1 \end{bmatrix} \begin{bmatrix} 0 \\ 0 \\ V_k \cos k\omega_0 t \end{bmatrix} = \begin{bmatrix} V_k \cos k\omega_0 t \\ V_k \cos k\omega_0 t \\ V_k \cos k\omega_0 t \end{bmatrix}. \tag{2.23}$$

Table 2.4 lists the effect of inverse transformations, where *dq*-frame signals are mapped into harmonics with different sequences. In particular, we should take care of *dq*-frame signals, which determine the orders of harmonics after inverse transformations.

In conclusion, we should pay special attention to the sequence of harmonics in imbalanced systems. To handle imbalanced voltages, we first expand them through Fourier series. Second, we decompose each dc, fundamental, and harmonic components into positive-sequence, negative-sequence, and zero-sequence components. Finally, we change the orders of components through frame transformations according to their sequences.

So far, we have discussed typical voltage quality problems, including amplitude deviations, harmonics, and imbalances. Utilities or power system operators must guarantee satisfactory voltage quality, thus supplying electric customers with standard grid voltages. If the majority of electric customers faces poor voltage quality, utilities should definitely be blamed. However, as electric loads may contaminate power grids, voltage quality problems in local areas are often caused by customer equipment, which affects grid voltages via current quality problems.

2.2 Current Quality Problems

In this section, we focus on current quality problems. Notably, current quality problems refer to the problems that are often caused by electric loads or customers. In particular, electric loads are supplied by standard grid voltages (i.e., with rated amplitudes, perfectly sinusoidal waveforms, and three-phase balance), they may still inject (or absorb) intolerable currents to (or from) power grids. Major current quality problems include reactive currents, current harmonics, and imbalanced currents. We will start with reactive currents.

2.2.1 Reactive Currents

Reactive currents refer to the fundamental current components that are orthogonal to fundamental voltages. As discussed in the previous section, grid voltages may contain harmonics in addition to fundamental components. Notably, reactive currents correspond to fundamental voltages instead of voltage harmonics. To analyze reactive currents, we ignore voltage harmonics in this subsection.

As mentioned, fundamental voltages are transformed into constants in the d-axis through the *abc*/*dq*0 transformation. Similarly, we can transform fundamental currents into constants in the d-axis. However, voltage and current phasors are not necessarily overlap along the d-axis, indicating that the q-axis current component can be nonzero. To validate, we apply the *abc*/*dq*0 transformation to three-phase fundamental currents, which lag three-phase voltages by an angle of φ, yielding

$$\frac{2}{3}\begin{bmatrix} \cos\omega_o t & \cos(\omega_o t - \frac{2\pi}{3}) & \cos(\omega_o t + \frac{2\pi}{3}) \\ -\sin\omega_o t & -\sin(\omega_o t - \frac{2\pi}{3}) & -\sin(\omega_o t + \frac{2\pi}{3}) \\ \frac{1}{2} & \frac{1}{2} & \frac{1}{2} \end{bmatrix} \begin{bmatrix} I_1\cos(\omega_o t - \varphi) \\ I_1\cos(\omega_o t - \frac{2\pi}{3} - \varphi) \\ I_1\cos(\omega_o t + \frac{2\pi}{3} - \varphi) \end{bmatrix} =$$

$$= \begin{bmatrix} I_1\cos\varphi \\ -I_1\sin\varphi \\ 0 \end{bmatrix} = \begin{bmatrix} i_{1d} \\ i_{1q} \\ 0 \end{bmatrix}, \tag{2.24}$$

where I_1 stands for the amplitude of fundamental currents, and i_{1d} and i_{1q} denote the d- and q-axis current components, respectively. Notably, we define the q-axis current component i_{1q} as the reactive current. If fundamental voltages and currents are in alignment (i.e., $\varphi = 0°$), $i_{1q} = 0$ is satisfied. Alternatively, we obtain the maximum absolute value of i_{1q} as I_1, provided that reactive currents are orthogonal to fundamental voltages (i.e., $\varphi = \pm 90°$). Figure 2.6 illustrates the relationship between the reactive current and voltage phasor, where the reactive current is negative in this case.

With the help of reactive currents, we can further calculate the active and reactive power through the following equations:

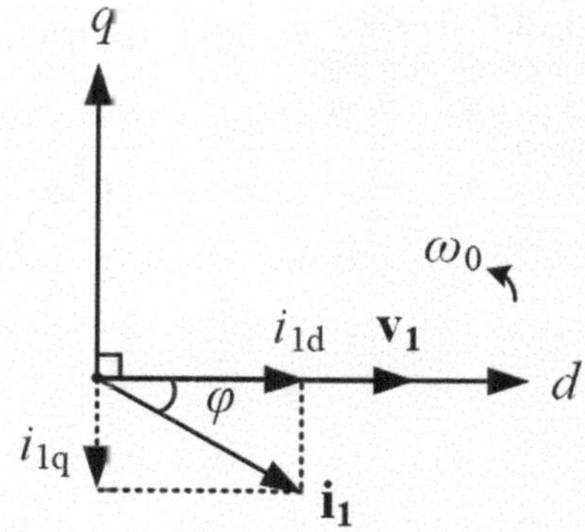

Fig. 2.6 Relationship between the reactive current and voltage phasor

$$p_1 = \frac{3}{2}(v_{1d}i_{1d} + v_{1q}i_{1q}) \text{ and} \tag{2.25}$$

$$q_1 = \frac{3}{2}(v_{1q}i_{1d} - v_{1d}i_{1q}), \tag{2.26}$$

where p_1 and q_1 represent the active and reactive power that are contributed by fundamental voltages and currents, respectively. It should be commented that voltage and current harmonics are ignored here. Otherwise, the interaction between voltage and current harmonics may change active and/or reactive power.

Moreover, the *complex power* $\mathbf{s}_1$ is calculated through

$$\mathbf{s}_1 = p_1 + jq_1. \tag{2.27}$$

In single-phase systems, the *average power* (or *active power*) and reactive power are defined as [4]

$$P_1 = \frac{1}{2}V_1 I_1 \cos\varphi \text{ and} \tag{2.28}$$

$$Q_1 = \frac{1}{2}V_1 I_1 \sin\varphi, \tag{2.29}$$

where V_1 and I_1 are the amplitudes of the fundamental voltage and current, respectively. Further, the complex power is expressed as

$$\mathbf{S}_1 = P_1 + jQ_1, \tag{2.30}$$

whose magnitude $|\mathbf{S}_1|$ is defined as the *apparent power*. Notably, we determine the power rating of electric equipment using its apparent power. As Q_1 contributes to $|\mathbf{S}_1|$, reactive power can increase power ratings. Therefore, it is desirable to reduce or remove reactive currents (or reactive power).

In addition, we define $\cos(\varphi)$ as the *power factor* i.e. [4],

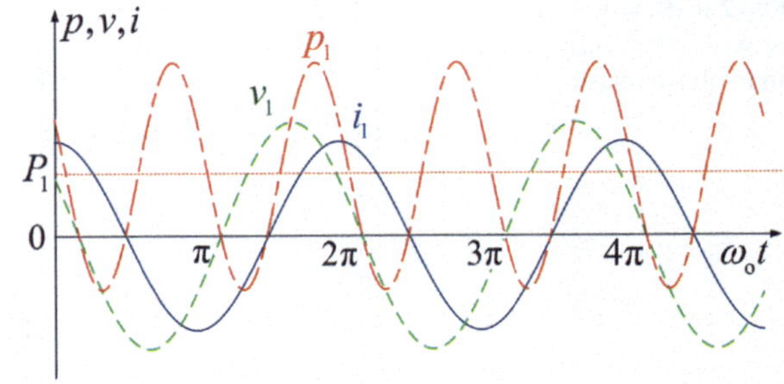

Fig. 2.7 Time-domain waveforms of voltage, current, and instantaneous power

$$\mathrm{PF} = \cos\varphi, \tag{2.31}$$

which equals one (or zero) when currents and voltages are in phase (or out-off phase). Normally, utilities prescribe the minimum power factor for electric loads (say 0.95) to reduce reactive currents. Correspondingly, φ is defined as the *power factor angle* [4]. Noticeably, a power factor angle ranges from 0 to 90° implies inductive loads. Alternatively, capacitive loads feature a power factor angle from 0 to −90°. In particular, $\varphi = 0°$ holds for resistive loads. In addition, the product of voltage and current signals in the time domain refers to the *instantaneous power*:

$$p_1(t) = v_1(t)i_1(t). \tag{2.32}$$

Figure 2.7 presents the time-domain waveforms of voltage, current, and instantaneous power, where the power factor angle φ is greater than 0° and less than 90°, thereby representing an inductive load. Obviously, the instantaneous power fluctuates at a double-line frequency. The instantaneous power can be negative due to the effect of reactive power. Techniques of improving power factors will be detailed in the next chapter.

2.2.2 *Current Harmonics*

In the previous section, we have detailed voltage harmonics. Similar definitions (such as THDs) are also applicable to current harmonics. When propagating along power lines, current harmonics often introduce voltage harmonics. Therefore, current harmonics should properly be treated.

Recapping Fig. 2.3, we have mentioned that diode rectifiers are harmonic sources. To validate, we supply single-phase and three-phase diode rectifiers with standard grid voltages and dc resistive loads. Next, we record simulated waveforms in Figs. 2.8 and 2.9, respectively. Clearly, voltage waveforms are highly sinusoidal and symmetrical. However, current waveforms are distorted due to the nonlinear nature of diode rectifiers. It is concluded from current waveforms in Figs. 2.8 and 2.9 that currents contain considerable amounts of harmonics.

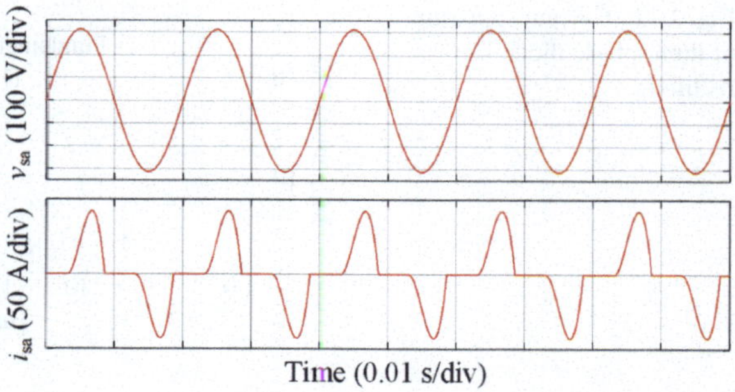

Fig. 2.8 Simulation waveforms of single-phase diode rectifiers

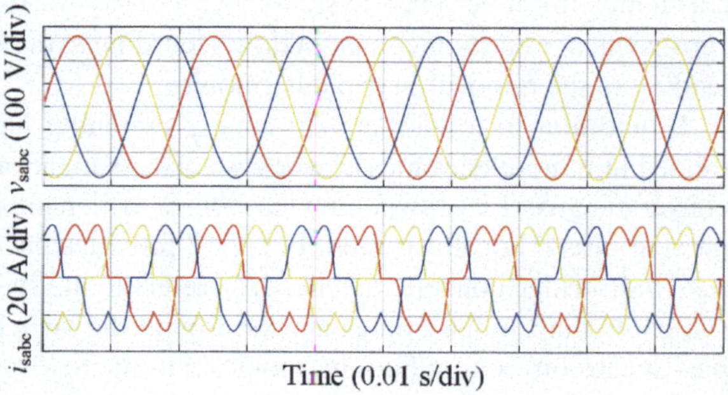

Fig. 2.9 Simulation waveforms of three-phase diode rectifiers

To quantitatively analyze harmonics, we calculate the current spectra of single-phase and three-phase diode rectifiers through fast-Fourier transform (FFT) analysis. The resultant spectra are shown in Figs. 2.10 and 2.11. As observed, single-phase diode rectifiers feature odd-order harmonics. In contrast, three-phase diode rectifiers feature $6n \pm 1$ harmonics (where n denotes a positive integer). Another interesting observation is that the amplitude of harmonics decreases as the harmonic order increases. This is due to the damping effect of filter inductors, as inductors exhibit high impedances in high-frequency domains. In comparison, harmonics in three-phase rectifiers are less conspicuous, as validated by THD comparisons (27.03% in the three-phase case vs. 63.51% in the single-phase case).

Current harmonics caused by diode rectifiers are representative. Dominant current harmonics are of a relatively low order, as the magnitude of high-order harmonics

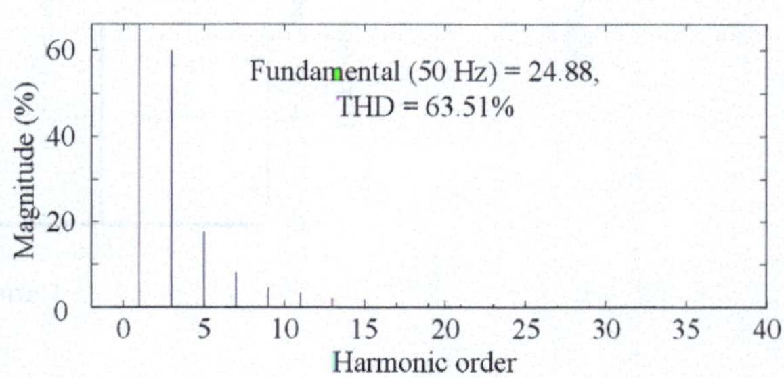

Fig. 2.10 Current spectrum of single-phase diode rectifiers

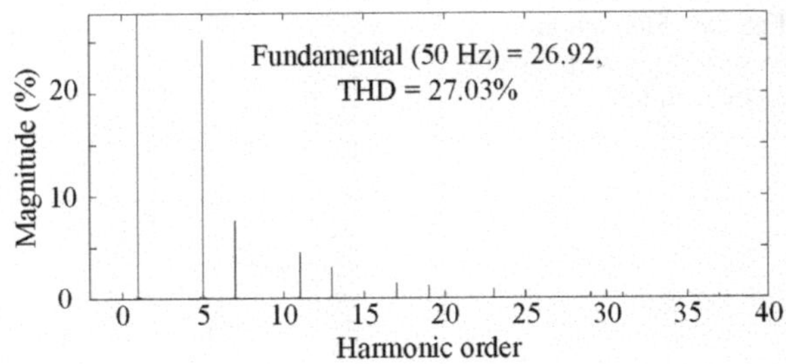

Fig. 2.11 Current spectrum of three-phase diode rectifiers

greatly reduces. In single-phase systems, odd-order harmonics outweigh even-order harmonics. In three-phase systems, particularly for three-phase three-line systems, zero-sequence harmonics cannot flow according to the Kirchhoff current law, thereby leading to the removal of triple harmonics.

In addition to low-order harmonics, switching harmonics should seriously be treated in more-electronics power systems. As mentioned, the modulation of power converters give rise to switching harmonics, which must sufficiently be attenuated by passive filters. For illustration, Fig. 2.12 provides the current spectrum of symmetrical half-bridge converters, where f_{sw} refers to the switching frequency. Obviously, switching current harmonics appear around the switching frequency and its multiples due to converter voltage harmonics. Similar to low-frequency current harmonics, switching current harmonics become weaker as the frequency increases. Once again, this is because of the increased impedance of passive filters in high-frequency domains.

Current inter-harmonics may appear in the case of nonlinear loads. However, harmonics are generally much stronger than inter-harmonics. We will discuss the solutions of current harmonics and inter-harmonics in the next chapter.

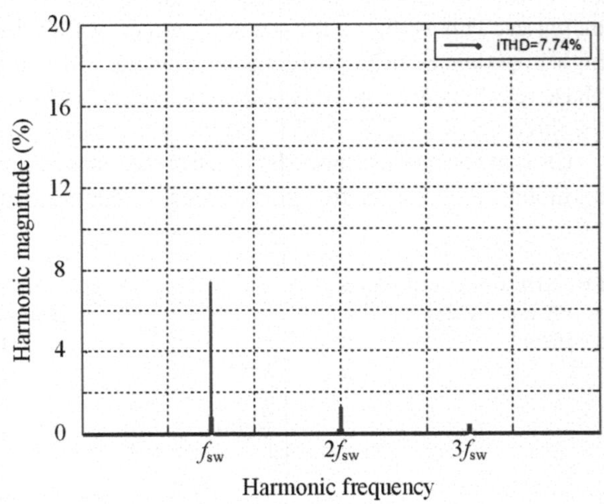

Fig. 2.12 Current spectrum of symmetrical half-bridge converters

2.2.3 Imbalanced Currents

Imbalanced currents are induced by either unsymmetrical loads or voltage imbalances. Generally, imbalanced voltages are translated into imbalanced currents with a significant amplification. In addition, unevenly distributed single-phase loads in three-phase systems and/or grid faults (e.g., single-phase short circuits) may cause imbalanced currents.

Also, sequence decomposition is applicable to imbalanced currents, leading to positive-sequence, negative-sequence, and zero-sequence currents. When calculating active power, we focus on the inner product of fundamental and harmonic components with the same sequence. Interactions among different sequences or harmonics are often ignored. However, they do influence equipment power ratings. Therefore, definitions of active, reactive, and complex power with imbalanced voltages and currents are still under active investigation and discussion.

Figure 2.13 demonstrates the experimental waveforms of balanced grid voltages and imbalanced converter currents, where the amplitude of the phase A current is larger than those of other phases, indicating a circuit imbalance. Such an imbalance is mainly caused by unsymmetrical converter loads.

In summary, voltage quality problems mainly comprise voltage amplitude deviations, voltage harmonics, and imbalanced voltages, while current quality problems include reactive currents, current harmonics, and imbalanced currents. Grid codes set standards on voltage and current quality. Now, we look at the relevant power quality standards.

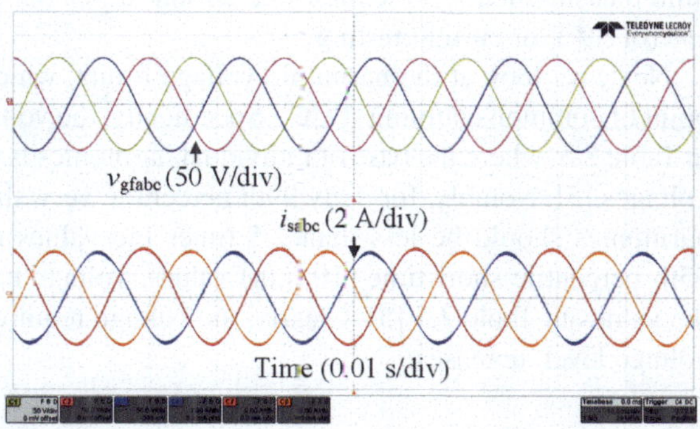

Fig. 2.13 Experimental waveforms of balanced grid voltages and imbalanced currents

2.3 Power Quality Standards

Institute of Electrical and Electronics Engineers (IEEE) and International Electrotechnical Commission (IEC) are two famous organizations who publish standards.

Voltage Amplitude

Regarding voltage amplitude deviations, within $\pm 10\%$ is a standard option. To protect customer equipment, some utilities lower the upper limit, e.g., to $+7\%$. Normally, SGs often generate higher voltages (e.g., $+5\%$) at their terminals to compensate voltage drops across transformers and transmission lines. Therefore, standard voltages are difficult to obtain at every node. In more-electronics power systems, power converters can regulate voltages well with minor amplitude errors (e.g., $\pm 1\%$). As a result, we can boost voltage quality in terms of amplitude deviations.

Voltage Harmonics

Before introducing standards on voltage harmonics, we first determine how to measure harmonics. According to IEEE Std. 519–2014, the harmonic measurement window (for FFT analysis) should feature a width of 10 fundamental cycles (i.e., 200 ms) for 50 Hz systems (or 12 cycles for 60 Hz systems) [5]. We aggregate 15 consecutive windows over a 3 s interval for very-short-time harmonic measurements [5]. Further, we aggregate 200 consecutive very-short-time values for 10 min short-time measurements [5]. On top of that, we should accumulate very-short- and short-time harmonic values for one day and one week, respectively [5]. For very-short-time measurements, we use the strongest 1% harmonic for testing. For short-time measurements, we exploit the strongest 5% harmonic for voltage testing and 1% (or 5%) for current testing [5].

Next, we look at the harmonic voltage limits, which refer to the voltages at the point of common coupling (PCC). Specifically, the voltage distortion limits are listed in Table 2.5, where the base of individual harmonics is the amplitude of fundamental voltages [5]. Notably, for daily 99th percentile very-short-time (3 s) values, voltage harmonics should be less than 1.5 times the values in Table 2.5 [5]. For weekly 95th percentile short-time (10 min) values, voltage harmonics should be less than the values in Table 2.5 [5]. Clearly, the voltage requirement becomes stricter as the voltage level increases.

Table 2.5 Voltage distortion limits [5]

Bus voltage V at PCC	Individual harmonic (%)	THD (%)
$V \leq 1.0$ kV	5.0	8.0
$1.0\text{ kV} < V \leq 69$ kV	3.0	5.0
$69\text{ kV} < V \leq 161$ kV	1.5	2.5
$161\text{ kV} < V$	1.0	1.5[a]

[a] High-voltage systems can have up to 2.0% THD where the cause is an HVDC terminal

Concerning voltage inter-harmonics, we have some draft standards. However, these standards have not received a widespread agreement.

Voltage Imbalances
Requirements on voltage imbalances are case dependent. As mentioned, it is generally required that voltage imbalances should be within the range of 1–3%. However, as voltage imbalances may amplify current imbalances, it is desirable to sufficiently limit voltage imbalances.

Reactive Currents
Requirements on reactive currents are often reflected by those on power factors. Normally, power factors are subscribed concerning electric loads. However, the nature of loads varies greatly. As a result, power factors change on a case-by-case basis. Generally, we put more stringent requirements on high-power loads. Typical power factor requirements are 0.95 or 0.9.

Current Harmonics
According to IEEE Std. 519–2014, standards on current distortion limits differ for systems with different voltage levels. For systems rated from 120 V to 69 kV, the relevant current distortion limits are documented in Table 2.5, where the total demand distortion (TDD) is defined as the square root of the sum of harmonic components divided by the maximum load fundamental current I_L [5]. It equals the THD under maximum load conditions. I_L is measured at the PCC and should be the sum of the currents according to the maximum demand and averaged over 12 months [5]. Noticeably, for daily 99th percentile very-short-time (3 s) values, current harmonics should be less than 2.0 times the values in Table 2.6. In other case, for weekly 99th percentile short-time (10 min) values, current harmonics should be less than 1.5 times the values in Table 2.6 [5]. Otherwise, for weekly 95th percentile short-time (10 min) values, current harmonics should be less than the values in Table 2.6 [5]. For higher voltage systems, current distortion limits become stricter.

It is noted from Table 2.6 that requirements become stricter as the harmonic order increases. This is understandable, as high-order harmonics feature smaller magnitudes and are attenuated better by passive filters. Moreover, more stringent standards are prescribed in *weak power grids*, which are characterized by a small ratio of I_{SC}/I_L. This is because weak grids feature larger grid impedances, and hence more significant harmonic voltage drops.

We should pay attention to the limits on switching harmonics, whose frequencies can go beyond that of 50th harmonic. Generally, switching harmonics should be attenuated below 0.3% I_L. In addition, similar to voltage inter-harmonics, current inter-harmonics have not been required in a unified standard.

Current Imbalances
Requirements on voltage imbalances are often used instead of current imbalances. This is because voltage imbalances may greatly amplify current imbalances.

Table 2.6 Current distortion limits for systems rated 120 V through 69 kV [5]

Maximum harmonic current distortion in percent of I_L						
Individual harmonic order (odd harmonics)[a,b]						
I_{SC}/I_L	$3 \le h < 11$	$11 \le h < 17$	$17 \le h < 23$	$23 \le h < 35$	$35 \le h < 50$	TDD
<20[c]	4.0	2.0	1.5	0.6	0.3	5.0
20 < 50	7.0	3.5	2.5	1.0	0.5	8.0
50 < 100	10.0	4.5	4.0	1.5	0.7	12.0
100 < 1000	12.0	5.5	5.0	2.0	1.0	15.0
>1000	15.0	7.0	6.0	2.5	1.4	20.0

[a]Even harmonics are limited to 25% of the odd harmonic limits above
[b]Current distortions that result in a DC offset, e.g., half-wave converters, are not allowed
[c]All power generation equipment is limited to these values of current distortion, regardless of actual I_{SC}/I_L
where I_{SC} = maximum short-circuit current at PCC
I_L = maximum demand load current (fundamental frequency component)
at the PCC under normal load operating conditions

In this chapter, we have discussed power quality problems in more-electronics power systems. As discussed, power quality problems are categorized into voltage quality problems and current quality problems. Voltage quality problems mainly include voltage amplitude deviations, voltage harmonics, and voltage imbalances. In contrast, current quality problems comprise reactive currents, current harmonics, and current imbalances. We have presented features and mechanisms of power quality problems as well as their standards.

References

1. Dugan RC, McGranaghan MF, Santoso S, Beaty HW (2003) Electrical power systems quality. McGraw-Hill, 2 Penn Plaza, New York, NY, USA
2. Erickson RW, Maksimovic D (2001) Fundamentals of power electronics. Springer, New York, NY, USA
3. Holmes DG, Lipo TA (2003) Pulse width modulation for power converters: principles and practive. Wiley, Hoboken, NJ
4. Nilsson JW, Riedel SA (2015) Electric circuits, 10th edn. Prentice Hall, New Jersey, NJ, USA
5. IEEE recommended practices and requirements for harmonic control in electric power systems, IEEE Standard 519–2014, 2014

Chapter 3
Power Quality Conditioning

3.1 Passive Solutions

Solutions to power quality problems are briefly classified into passive and active solutions. Passive solutions employ passive components for power quality enhancement. Alternatively, active solutions involve power electronic devices or equipment. In more-electronics power systems, active solutions play an increasingly important role on power quality conditioning.

3.1.1 *Voltage Amplitude and Reactive Power Compensation*

As mentioned, voltages sags are the most common power quality problems. Normally, voltage drops across inductive impedances cause voltage sags. In consequence, capacitive components can compensate voltage sags.

To illustrate this statement, Fig. 3.1 shows a phasor diagram of voltage sags, where $\mathbf{v}_{\mathbf{grid}}$ and $\mathbf{v}_{\mathbf{load}}$ stand for the standard grid voltage and load voltage phasors, respectively. Additionally, the inductive current phasor $\mathbf{i}_{\mathbf{load}}$ lags the two voltage phasors by 90°. As mentioned in Sect. 1.3.3, the simplest model of transmission lines is an inductor, whose impedance is assumed to be jX_{line}. Correspondingly, the voltage drop across this inductor is $jX_{\text{line}}\mathbf{i}_{\mathbf{load}}$, which is in alignment with $\mathbf{v}_{\mathbf{grid}}$. As a result, the magnitude of the load voltage $|\mathbf{v}_{\mathbf{load}}|$ reduces as compared to $|\mathbf{v}_{\mathbf{grid}}|$, indicating that voltage sags occur.

There are two options to implement capacitive compensation. First, we compensate line impedances so that voltage drops become less prominent. Second, we compensate inductive loads, leading to the improvements of not only voltage amplitudes but also power factors. In the second case, voltage amplitude and reactive power (or currents) compensation can simultaneously be achieved.

J. Fang, *More-Electronics Power Systems: Power Quality and Stability*,
Power Systems, https://doi.org/10.1007/978-981-15-8590-6_3

Figure 3.2 visualizes these two capacitive compensation options, where X_{line}^{*} and $\mathbf{i}_{\mathbf{load}}^{*}$ represent the compensated line impedance and load current phasor, respectively. After compensation, the magnitudes of the load voltage $|\mathbf{v}_{\mathbf{load}}|$ increases in both cases as compared with that in Fig. 3.1. Notably, the change of X_{line}^{*} is under the assumption of a fixed load current, and vice versa.

Correspondingly, we connect capacitors with power grids either in parallel or in series. In comparison, paralleled capacitors are more popular due to their flexibility and effectiveness, particularly for distribution systems. In particular, switched-capacitor banks, consisting of paralleled switched-capacitors, are increasingly used for finer voltage and reactive power control. Figure 3.3 shows a schematic of switched-capacitor banks, where switches can be either mechanical switches or thyristors. The latter case is known as thyristor-switched capacitors (TSCs) [1].

Dually, inductors serve to compensate voltage sags and reactive power in the case of underground cables, which exhibit capacitive impedances, or voltage swells under the condition of overhead lines. Similar to capacitors, we connect inductors in parallel with power grids. When controlled by thyristors, they are named as thyristors-controlled reactors (TCRs) [1]. TSCs and TCRs (among others) are collectively referred to as static var compensators (SVCs). The keyword "static" is to differentiate SVCs from rotational SGs or synchronous condensers, which are essentially SGs without prime movers for reactive power and voltage regulation.

Notably, we employ capacitors and inductors simultaneously in static var systems, as shown in Fig. 3.4, where the series LC branch targets at harmonic filtering, which will be discussed in the next subsection.

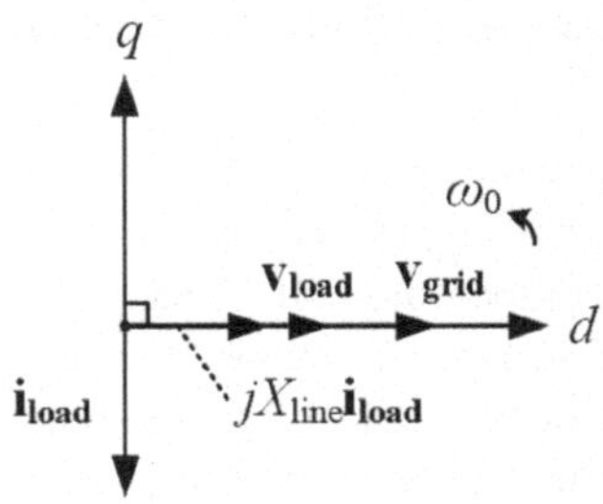

Fig. 3.1 Phasor diagram of voltage sags

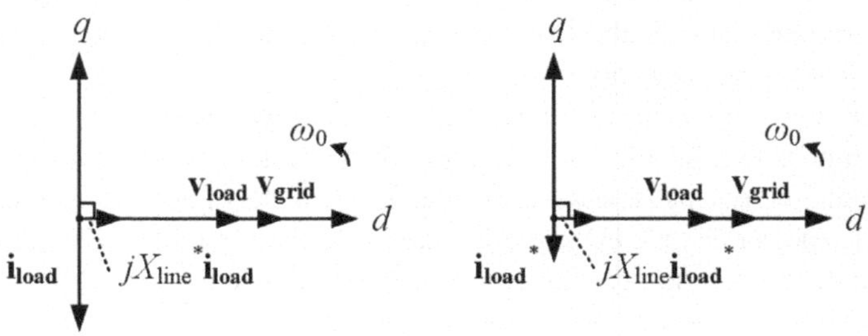

Fig. 3.2 Capacitive compensation of line impedances and loads

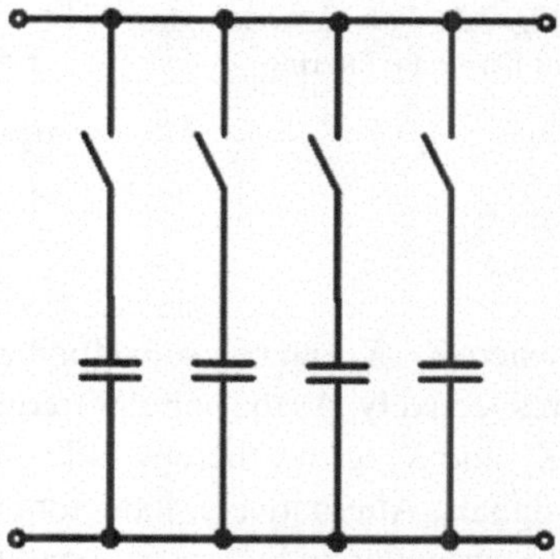

Fig. 3.3 Switched-capacitor banks

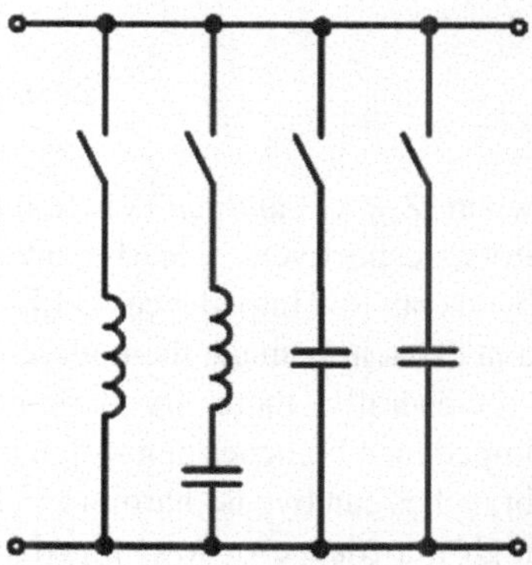

Fig. 3.4 Schematic of static var systems

In general, passive solutions for voltage and reactive power compensation are simple and mature. However, the solutions are not very flexible due to discontinuous compensation. In addition, capacitors and inductors are bulky and expensive components with potential resonance hazards. As will be introduced, power-electronic-based active solutions get rid of such drawbacks, thus becoming popular in recent years. Before going to active solutions, let us discuss passive filters.

3.1.2 Harmonic Filtering

Passive filters are essential for harmonic filtering, particularly for switching harmonic filtering. Basic passive filtering components mainly include inductors and capacitors. First, we recap their impedance features.

It is known that inductors feature an impedance of [2]

$$Z_{\mathrm{L}}(\omega) = jX_{\mathrm{L}}(\omega) = j\omega L_{\mathrm{L}}, \tag{3.1}$$

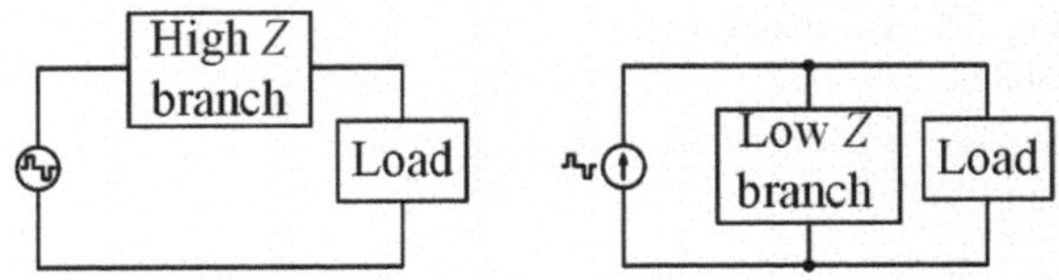

Fig. 3.5 Two methodologies of harmonic filtering

where Z_L, X_L, and L_L stand for the impedance, reactance, and inductance of inductors, respectively. As the angular frequency (or frequency) ω increases, the magnitudes of Z_L and X_L grow, thereby indicating that inductors tend to suppress high-frequency signals. Alternatively, inductors feature small impedances in low-frequency bands. In particular, inductors are conductors for dc signals.

Alternatively, the impedance of capacitors is expressed as [2]

$$Z_C(\omega) = -jX_C(\omega) = \frac{-j}{\omega C_C}, \tag{3.2}$$

where Z_C, X_C, and C_C denote the impedance, reactance, and capacitance of capacitors, respectively. Clearly, capacitors exhibit high impedances in low-frequency bands but low impedances at high frequencies. Complimentary features of inductors and capacitors make them ideal partners for harmonic filtering.

Generally, there are two methodologies for harmonic filtering. First, high-impedance branches can stop harmonics from propagation. Second, low-impedance branches can bypass harmonics, leading to clean signals. It is worth mentioning that high-impedance or low-impedance branches are suitable for harmonic voltage or current sources, respectively. According to the circuit theory [2], every branch that is paralleled with a voltage source can be equivalent to a voltage source. Therefore, we employ high-impedance branches in series with voltage sources. Alternatively, low-impedance branches are suitable for current sources. Figure 3.5 visualizes the two methodologies of harmonic filtering.

3.1.2.1 Low-Frequency Harmonic Filtering

Low-frequency harmonics refer to dominant harmonics introduced by nonlinear loads, such as diode rectifiers. Conventionally, passive filters are employed to filter low-frequency harmonics.

As discussed, capacitors or inductors feature low impedances in high-frequency or low-frequency bands, respectively. To filter low-frequency harmonics, series capacitors and/or parallel inductors are suitable. In addition to being used alone, they can lump into passive filters.

For dc signal filtering, we rarely use parallel inductors due to size and loss drawbacks. Instead, series capacitors are well known approaches of isolating dc signals in analog circuits [2]. When simultaneously employed, inductors and capacitors become *LC* filters, as shown in Fig. 3.6. In the left-hand part of Fig. 3.6, we notice a series

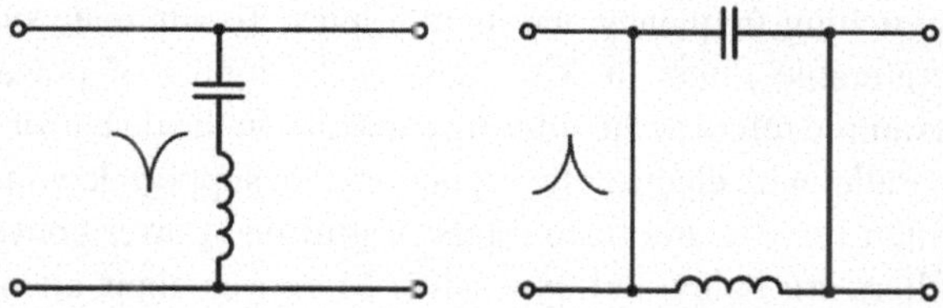

Fig. 3.6 Schematics of *LC* filters

LC filter. Its impedance takes the form of

$$Z_{\text{LC\_series}}(\omega) = jX_{\text{L}}(\omega) - jX_{\text{C}}(\omega) = j(\omega L_{\text{L}} - \frac{1}{\omega C_{\text{C}}}), \tag{3.3}$$

whose amplitude follows a notch shape, as depicted in the left-hand side of Fig. 3.6. In particular, the impedance of series *LC* filters is zero at the resonance frequency (or resonance angular frequency). By setting (3.3) to zero, we obtain

$$Z_{\text{LC\_series}}(\omega) = 0 \Rightarrow f_{\text{r\_LC}} = \frac{1}{2\pi\sqrt{L_{\text{L}} C_{\text{C}}}}. \tag{3.4}$$

In contrast, the impedance of paralleled *LC* filters is

$$Z_{\text{LC\_parallel}}(\omega) = jX_{\text{L}}(\omega)//jX_{\text{C}}(\omega) = \frac{j\omega L_{\text{L}}}{(1 - \omega^2 L_{\text{L}} C_{\text{C}})}, \tag{3.5}$$

which performs a band-stop function (see Fig. 3.6). The paralleled impedance becomes infinitely large at the resonance frequency

$$Z_{\text{LC\_parallel}}(\omega) = \infty \Rightarrow f_{\text{r\_LC}} = \frac{1}{2\pi\sqrt{L_{\text{L}} C_{\text{C}}}}. \tag{3.6}$$

By tuning the resonance frequency of *LC* filters at a dominant harmonic frequency (e.g., 150 Hz for single-phase 50 Hz systems or 300 Hz for three-phase 60 Hz systems), we achieve effective harmonic filtering. In practice, series *LC* filters are often used in combination with capacitor banks, forming a static var system (see Fig. 3.4). It should be reminded that real *LC* components have equivalent series resistors (ESRs), which damp resonant peaks to a certain extent.

3.1.2.2 Switching Harmonic Filtering

In high-frequency domains, inductors (or capacitors) feature high (or low) impedances. As such, we can employ either series inductors or paralleled capacitors for switching harmonic filtering according to Fig. 3.5.

As mentioned in the previous chapter (in Sect. 2.1), the modulation of power converters introduces voltage switching harmonics, which often appear around the switching frequency and its multiples. To attenuate such harmonics, capacitors are preferable filters in ICs, such as the inputs of power supplies. Another notable example refers to the filtering capacitor in the dc bus of power converters, which is in parallel with electrolytic capacitors that support dc voltages. In the ac side, inductors often serve as interface filters of grid-tied power converters [3]. However, inductor filters are bulky and expensive. Moreover, their attenuation around the switching frequency and its multiples is limited, as proved by Fig. 2.12. Thus, it is highly desirable to improve filtering performances with smaller and cheaper filters.

LCL Filters

LCL filters are popular third-order filters for switching (voltage) harmonic filtering. As compared with single inductor *L* filters, *LCL* filters feature smaller inductances (and hence lower voltage drops and higher control bandwidths), lower costs, smaller size, and better filtering. Figure 3.7 shows the schematic of *LCL* filters, which intend to link two voltage sources (e.g., converter-voltage and grid-voltage sources). Notably, the filter capacitor C_{cf} can bypass high-frequency converter harmonics, leading to clear grid-injected currents. In the case of current sources, the *LCL* filter should be replaced by its dual filter—the *π filter* (or the *CLC filter*). Without loss of generality, we focus on the condition of voltage sources.

According to Table 2.6, switching current harmonics should be attenuated to be less than the 0.3% maximum demand load current. Besides, the rated grid current THD must be less than 5%. To evaluate harmonic attenuation, we derive the transfer function from the converter voltage to the grid-injected current from Fig. 3.7 as

$$\frac{\Delta i_{\mathrm{cg}}(s)}{\Delta v_{\mathrm{ci}}(s)} = \frac{Z_{\mathrm{cf}}(s)}{Z_{\mathrm{cg}}(s)Z_{\mathrm{cf}}(s) + Z_{\mathrm{ci}}(s)Z_{\mathrm{cg}}(s) + Z_{\mathrm{ci}}(s)Z_{\mathrm{cf}}(s)}, \tag{3.7}$$

where $Z_{\mathrm{ci}}(s)$, $Z_{\mathrm{cf}}(s)$, and $Z_{\mathrm{cg}}(s)$ represent the impedances of respective branches. Substitution of $Z_{\mathrm{ci}}(s) = L_{\mathrm{ci}}s$, $Z_{\mathrm{cg}}(s) = L_{\mathrm{cg}}s$, and $Z_{\mathrm{cf}}(s) = 1/(C_{\mathrm{cf}}s)$ into (3.7), it yields the transfer function of *LCL* filters:

$$G_{\mathrm{LCL}}(s) = \frac{\Delta i_{\mathrm{cg}}(s)}{\Delta v_{\mathrm{ci}}(s)} = \frac{1}{L_{\mathrm{cg}}L_{\mathrm{ci}}C_{\mathrm{cf}}s^3 + (L_{\mathrm{cg}} + L_{\mathrm{ci}})s}. \tag{3.8}$$

Substituting of $s = j\omega$ into (3.8) and letting its denominator equal zero, we derive the resonance frequency of *LCL* filters as

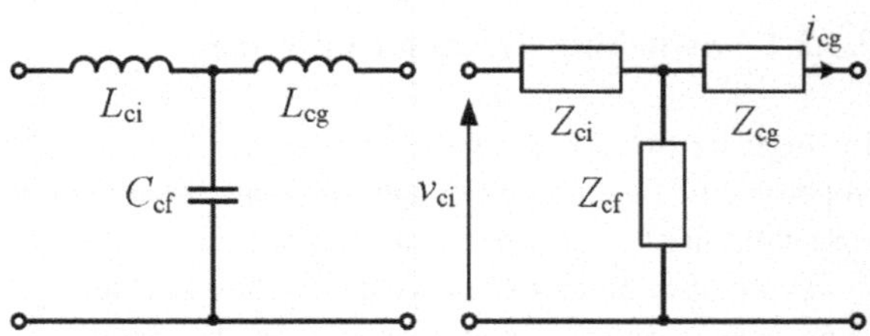

Fig. 3.7 Schematics of *LCL* filters and their general circuits

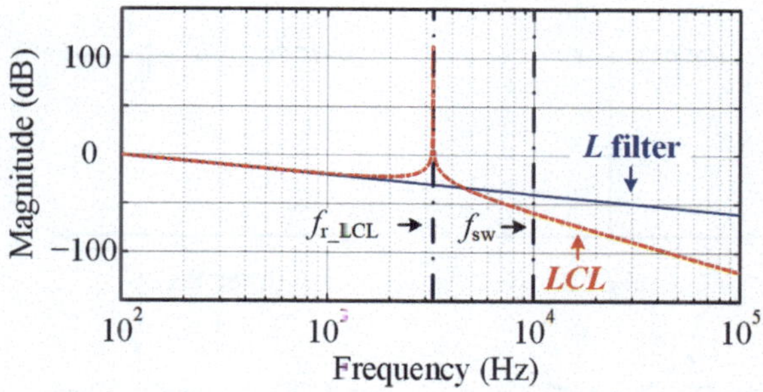

Fig. 3.8 Bode diagrams of the converter voltage to the grid current transfer functions of *L*- and *LCL*-filtered grid-tied power converters

$$f_{r\_LCL}(s) = \frac{1}{2\pi}\sqrt{\frac{L_{cg} + L_{ci}}{L_{cg}L_{ci}C_{cf}}}. \tag{3.9}$$

Figure 3.8 shows the Bode diagrams of the converter voltage to the grid current transfer functions of *L*- and *LCL*-filtered power converters, where f_{sw} denotes the switching frequency. Clearly, the *LCL* filter performs much stronger attenuation around f_{sw} due to its −60 dB/dec (decade) roll-off rate. In contrast, the *L* filter exhibits a roll-off rate of −20 dB/dec, thus resulting in more switching harmonics.

Despite with stronger attenuation than single *L* filters, *LCL* filters may still fail to comply with harmonic standards in demanding applications, such as half-bridge APFs, where high control bandwidths and strong attenuation ability are simultaneously required [4]. To further improve filter size, costs, and performances, modified *LCL* filters with additional *LC* traps have been proposed [5]. In addition, it should be highlighted that *LCL* filters introduce resonances (see Fig. 3.8), whose peaks should be properly attenuated for stable systems, as will be discussed.

Modified LCL Filters

Figure 3.9 illustrates the schematics of modified *LCL* power filters (including the *LCL* filter), where the model of line impedances is not shown for simplicity, and its effect will be incorporated wherever necessary [5]. Note that all the modified *LCL* filters contain series and/or paralleled *LC* traps.

As discussed, Fig. 3.9a shows the *LCL* filter, where two filter inductors and one filter capacitor are employed. In addition, Fig. 3.9b gives the series *LC* filter. This filter is mainly used in reactive power conditioning applications, where the large voltage drop across the filter capacitor reduces voltage and power ratings of grid-tied power converters [6]. On top of that, the series *LC* filter is often tuned at a low-order harmonic (e.g., the 7th harmonic) instead of switching harmonics. Therefore, the series *LC* filter only provides moderate attenuation to switching harmonics.

Figure 3.9c illustrates the schematic of *LLCL* filters [7]. By inserting a small trap inductor into the filter capacitor branch loop of the *LCL* filter, the *LLCL* filter possesses a trap tuned at the switching frequency, which allows a significant reduction of filter inductances while maintaining satisfactory harmonic attenuation [7].

According to the dual relationship between series and parallel circuits, a small trap capacitor is added in parallel with the grid-side inductor of an *LCL* filter, forming

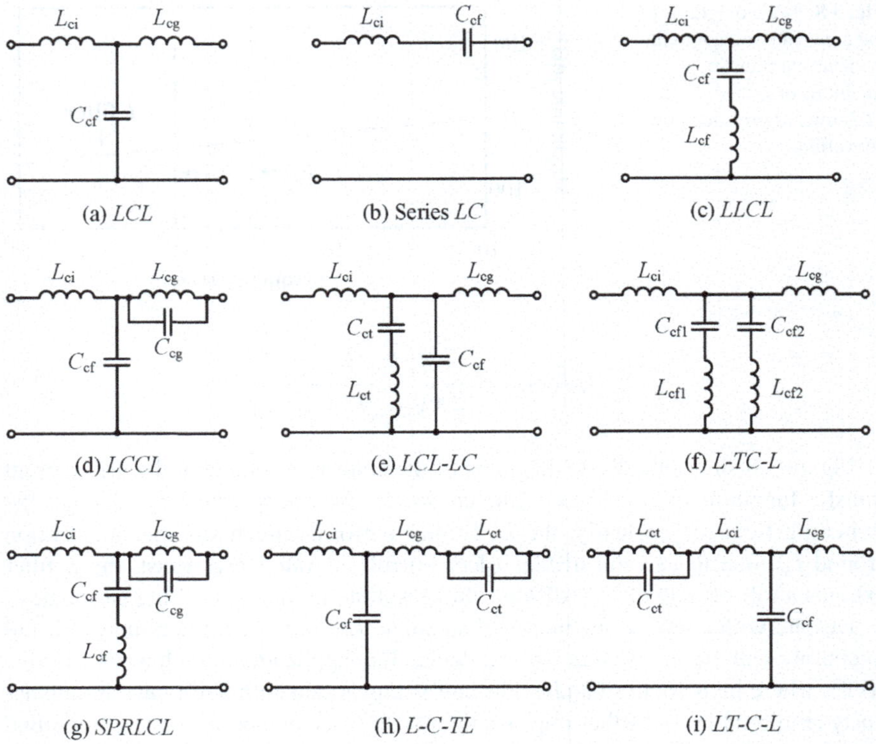

Fig. 3.9 Schematics of modified *LCL* filters [5]

the *LCCL* filter, as shown in Fig. 3.9d. Generally, trap capacitors are smaller and cheaper than trap inductors. As a result, *LCCL* filters are more cost-effective than *LLCL* filters [8].

Figure 3.10 shows the Bode diagrams of the converter voltage to the grid current transfer functions of the grid-tied power converters with modified *LCL* filters. Clearly, *LLCL* and *LCCL* filters perform strong attenuation around the switching

Fig. 3.10 Bode diagrams of the converter voltage to the grid current transfer functions of the grid-tied power converters with modified *LCL* filters

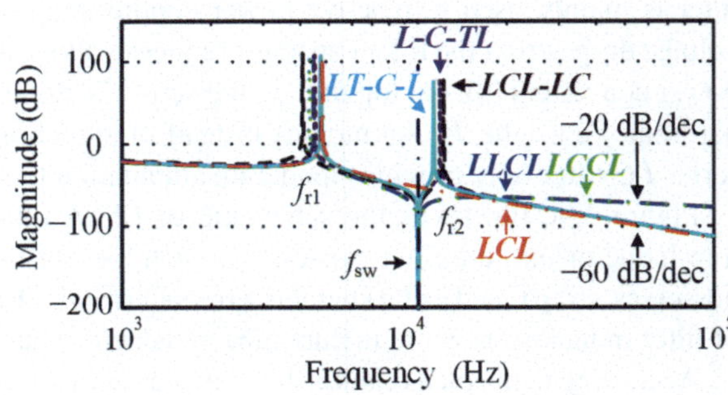

frequency f_{sw}. However, their high-frequency roll-off rates become worse, which are −20 dB/dec as *L* filters. Therefore, *LLCL* and *LCCL* filters exhibit poorer harmonic filtering at multiples of f_{sw} as compared with *LCL* filters.

To further improve switching harmonic filtering performances of *LLCL* filters, we incorporate an extra capacitor based on an *LLCL* filter to construct the *LCL-LC* filter, as depicted in Fig. 3.9e [9, 10]. Also, the *LCL-LC* filter can be regarded as a combination of one *LCL* filter and one series *LC* trap.

The *L-TC-L* filter in Fig. 3.9f contains more than one series *LC* traps [11–13]. These *LC* traps are tuned at the switching frequency and its multiples. Despite its strong attenuation, the *L-TC-L* filter is very complicated even with only two traps. It should be emphasized that the *L-TC-L* filter may contain another filter capacitor, which is similar to the case of *LCL-LC* filters, for improving high-frequency attenuation at the expense of filter complexity [11].

When combining the series-resonant trap of *LLCL* filters and parallel-resonant trap of *LCCL* filters, we derive the series–parallel-resonant *LCL* filter (*SPRLCL* filter), as illustrated in Fig. 3.9g [4]. As compared with the *L-TC-L* filter, the *SPRLCL* filter saves one trap inductor. However, the *SPRLCL* filter also suffers from a poor roll-off rate of −20 dB/dec.

Figure 3.9h presents the *L-C-TL* filter, which can be regarded as a combination of one *LCL* filter and one grid-side *LC* trap [14]. Also, the *L-C-TL* filter is identical to the *LCCL* filter when the line impedance is inductive. Referring to Fig. 3.10, we note that the *L-C-TL* filter possesses a desirable roll-off rate of −60 dB/dec.

When the *LC* trap is shifted from the grid side to the converter side, it is possible to derive the *LT-C-L* filter, as shown in Fig. 3.9i. As observed in Fig. 3.10, the *LT-C-L* filter performs strong attenuation in the high-frequency band and a trapping feature around the switching frequency.

Among all the passive filters in Fig. 3.9, the series *LC* filter is the simplest one with only two filter components. However, it is application specific without widespread applications. In contrast, the *LCL* filter containing three filter components eclipses the single inductor *L* filter in many applications, but it performs no harmonic trapping around the switching frequency. As fourth-order filters, *LLCL* and *LCCL* filters incorporate *LC* traps, hence featuring very strong attenuation around the switching frequency. The *LCL-LC*, *L-C-TL*, and *LT-C-L* filters are fifth-order filters evolved from *LLCL* and *LCCL* filters for better high-frequency harmonic attenuation. The *SPRLCL* and *L-TC-L* filters contain at least two *LC*-traps, and hence complicated filter structures. The respective filter orders are more than five and six, respectively.

In terms of roll-off rates, the *LCL-LC* filter, *L-C-TL* filter, and *LT-C-L* filter feature a −60 dB/dec roll-off rate as that of *LCL* filters, indicating their superior filtering performances. However, it is noted from Fig. 3.10 that new resonant peaks appear for these three passive filters. In practice, we should avoid designing these resonant peaks near the multiples of f_{sw}.

Table 3.1 summarizes the comparisons among passive filters. In particular, the robustness of filters focuses on the influence of grid inductances on switching harmonic trapping. In conclusion, the *LT-C-L* filter is claimed to be the most

Table 3.1 Comparisons of passive filters

Filter types	Additional components	Filter orders	Trap frequencies	Roll-off rates (dB/dec)	Robustness against *Ls*
LCL	–	3	–	−60	–
Series *LC*	–	2	Low-order harmonic	−20	N
LLCL	Trap inductor	4	f_{sw}	−20	Y
LCCL	Trap capacitor	4	f_{sw}	−20	Y
LCL-LC	*LC*-trap	5	f_{sw}	−60	Y
L-TC-L	*LC*-traps	≥6	f_{sw}, $2f_{sw}$, …	−20	Y
SPRLCL	Trap inductor and capacitor	≥s5	f_{sw} and $2f_{sw}$	−20	Y
L-C-TL	*LC*-trap	5	f_{sw}	−60	N
LT-C-L	*LC*-trap	5	f_{sw}	−60	Y

promising modified filter in terms of filter topology, filtering performance, and robustness [5].

3.1.2.3 Design of Passive Filters

After introducing various passive filters, we proceed to the design of passive filters in this part. Overall, the design of passive filters, particularly for high-order filters, is tricky. This can be evidenced by the hundreds of research papers on the design and analysis of *LCL* filters. As such, we focus on the main points here.

For passive filters targeting at low-order harmonics, design becomes simple. The key lies in the proper selection of resonance frequencies, which should be consistent with major harmonic frequencies. On top of that, passive filters may possibly be required to compensate a certain amount of reactive power, which in turn determines the filter impedance and parameters at the fundamental frequency.

For high-order filters targeting at switching harmonics, design is much more complicated. In this case, there are several factors to consider. These factors are often intertwined with each other in practice. As a result, tradeoffs among filtering performances, costs, size, stability margins, and complexity, are necessary. The following aspects should be taken into consideration.

Harmonic Standard

Passive filters should provide sufficient attenuation to switching harmonics so that they comply with grid codes (see Table 2.6). Specifically, the magnitude of the transfer functions from converter voltages to grid currents [e.g., in (3.8)] is an important index. Given operating conditions, we can derive switching voltage harmonics through formulas in Chap. 2 [15]. Next, according to Table 2.6, the required attenuation abilities at the switching frequency and its multiples can be derived. For passive filters without any LC trap, we evaluate their attenuation around the switching frequency. For trap filters, we evaluate the attenuation around multiples of the switching frequency, dependent on the number of LC traps, as detailed in [4]. The requirements of harmonic standards set an upper bound of resonance frequencies, which further depend on major filtering components.

Converter Ripple

To protect semiconductor switches and avoid magnetic saturation, it is necessary to limit converter ripples (e.g., within 40%). Similar requirements exist in the design of passive filters in dc-dc converters [16]. The ripple requirements put a minimum limit on converter inductances.

Reactive Power

Generally, high-order passive filters should not consume much reactive power. This translates into an upper limit on filtering capacitors. A practical limit on reactive power absorption by capacitors is 5% [17].

Size, Weight, and Cost

We have discussed these key features of power converters in Chap. 1. To reduce filter size, weight, and cost, we prefer filters with lower inductances and capacitances. In this sense, magnetic integrated inductors can be desirable [18].

Control Bandwidth and Stability

In terms of control bandwidth and stability, the resonance frequency of passive filters should be within certain ranges. For example, the control bandwidth of APFs should be wide enough for harmonic compensation. We will open up this point later.

3.2 Uninterruptable Power Supplies (UPSs)

We continue to explore active solutions of power quality problems, starting with uninterruptable power supplies (UPSs). As the name suggests, UPSs provide high-quality power supplies even when power grids are off. Generally, UPSs intend to address voltage quality problems, including voltage amplitude deviations, voltage harmonics, and/or voltage imbalances.

3.2.1 Fundamental Principles

UPSs are essentially dc-ac power converters that tightly regulate their ac voltages. Remember that topologies, modeling, and control are the three important subtopics of power converters, as detailed in Sect. 1.2. We look at these aspects of UPSs. In terms of topologies, UPSs feature standard inverter topologies, such as single-phase

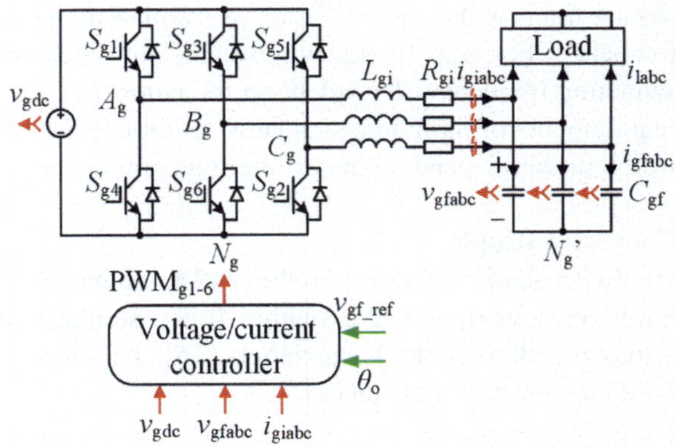

Fig. 3.11 Schematic of three-phase-three-line UPSs

half-bridge, single-phase H-bridge, three-phase-three-line half-bridge, three-phase-four-line half-bridge, three-phase full-bridge, or multilevel circuits.

For illustration, Fig. 3.11 shows the schematic of standard three-phase-three-line half-bridge UPSs, where the dc bus represents a battery output, whose voltage is denoted as v_{gdc}. The semiconductor switches S_{g1}–S_{g6} convert v_{gdc} into ac voltages v_{gfabc}. The fundamental control objective of UPSs is to make v_{gfabc} follow their references, including the voltage amplitude v_{gf_ref} and phase angle θ_0. As a result, critical loads are supplied by UPSs with high-quality power.

Notably, there are various types of UPSs. As one example, an UPS can continue to work, where the energy comes from batteries, which are supplied by diesel generators or power grids through rectifiers. In another example, an UPS operates in parallel to power grids. The UPS only contributes power when the power quality of grids falls below a certain limit. Under normal conditions, power grids support critical loads.

3.2.2 Plant Modeling

Next, we investigate the plant model of UPSs. Similar to modelling of dc-dc power converters, we first build differential equations using the Kirchhoff's circuit laws. Let v_{gxN} ($x = a, b, c$) represent the converter voltages across the nodes X_g ($X_g = A_g$, B_g, C_g) and N_g, respectively. According to the Kirchhoff's voltage law (KVL), we obtain the following equations:

$$\begin{cases} v_{gaN}(t) = v_{gfa}(t) + L_{gi}\frac{di_{gia}(t)}{dt} + R_{gi}i_{gia}(t) + v_{gN'N}(t) \\ v_{gbN}(t) = v_{gfb}(t) + L_{gi}\frac{di_{gib}(t)}{dt} + R_{gi}i_{gib}(t) + v_{gN'N}(t) \\ v_{gcN}(t) = v_{gfc}(t) + L_{gi}\frac{di_{gic}(t)}{dt} + R_{gi}i_{gic}(t) + v_{gN'N}(t) \end{cases}, \tag{3.10}$$

where $v_{gN'N}$ stands for the voltage between nodes $N_g{}'$ and N_g.

We express the capacitor voltages v_{gfabc} as functions of the filter currents i_{giabc} and load currents i_{labc} as

$$\begin{cases} v_{gfa}(t) = \frac{1}{C_{gf}} \int \left[i_{gia}(t) - i_{la}(t) \right] \\ v_{gfb}(t) = \frac{1}{C_{gf}} \int \left[i_{gib}(t) - i_{lb}(t) \right] \\ v_{gfc}(t) = \frac{1}{C_{gf}} \int \left[i_{gic}(t) - i_{lc}(t) \right] \end{cases}. \tag{3.11}$$

For three-phase three-wire systems, the sum of three-phase currents should be zero. By summing the three equations in (3.11), we find that the sum of three-phase capacitor voltages equals zero. With this, we sum the equations in (3.10), yielding

$$v_{gN'N}(t) = \frac{v_{gaN}(t) + v_{gbN}(t) + v_{gcN}(t)}{3}, \tag{3.12}$$

which upon substitution into (3.10) gives

$$\begin{cases} \frac{2v_{gaN}(t) - v_{gbN}(t) - v_{gcN}(t)}{3} = v_{gfa}(t) + L_{gi}\frac{di_{gia}(t)}{dt} + R_{gi}i_{gia}(t) \\ \frac{-v_{gaN}(t) + 2v_{gbN}(t) - v_{gcN}(t)}{3} = v_{gfb}(t) + L_{gi}\frac{di_{gib}(t)}{dt} + R_{gi}i_{gib}(t) \\ \frac{-v_{gaN}(t) - v_{gbN}(t) + 2v_{gcN}(t)}{3} = v_{gfc}(t) + L_{gi}\frac{di_{gic}(t)}{dt} + R_{gi}i_{gic}(t) \end{cases}. \tag{3.13}$$

We define S_{gabc} as switch functions. Specifically, the state of $S_{gabc} = 1$ corresponds to upper switches on and lower switches off, and vice versa in the case of $S_{gabc} = 0$. Following this definition, we have $v_{gabcN} = S_{gabc}v_{gdc}$, which after substituted into (3.13) leads to

$$\begin{cases} \frac{2S_{ga}(t) - S_{gb}(t) - S_{gc}(t)}{3} V_{gdc\_ref} = v_{gfa}(t) + L_{gi}\frac{di_{gia}(t)}{dt} + R_{gi}i_{gia}(t) \\ \frac{-S_{ga}(t) + 2S_{gb}(t) - S_{gc}(t)}{3} V_{gdc\_ref} = v_{gfb}(t) + L_{gi}\frac{di_{gib}(t)}{dt} + R_{gi}i_{gib}(t) \\ \frac{-S_{ga}(t) - S_{gb}(t) + 2S_{gc}(t)}{3} V_{gdc\_ref} = v_{gfc}(t) + L_{gi}\frac{di_{gic}(t)}{dt} + R_{gi}i_{gic}(t) \end{cases}, \tag{3.14}$$

where v_{gdc} is replaced by a constant V_{gdc_ref}, thereby removing the nonlinearity introduced by variable products. However, the model is still time variant, in terms of both switch functions as well as voltage and current signals.

To derive linear models, we replace natural-frame signals by their counterparts in the *dq*0-frame using the inverse transformation matrix $\mathbf{T_{dq0/abc}}$, leading to

$$\begin{cases} V_{gdc\_ref}S_{gd}(t) = v_{gfd}(t) + L_{gi}\frac{di_{gid}(t)}{dt} + R_{gi}i_{gid}(t) - \omega_o L_{gi}i_{giq}(t) \\ V_{gdc\_ref}S_{gq}(t) = v_{gfq}(t) + L_{gi}\frac{di_{giq}(t)}{dt} + R_{gi}i_{giq}(t) + \omega_o L_{gi}i_{gid}(t) \\ 0 = v_{gf0}(t) + L_{gi}\frac{di_{gi0}(t)}{dt} + R_{gi}i_{gi0}(t) \end{cases} \text{ and} \tag{3.15}$$

$$\begin{cases} C_{gf}\frac{dv_{gfd}(t)}{dt} - \omega_o C_{gf}v_{gfq}(t) = i_{gid}(t) - i_{ld}(t) \\ C_{gf}\frac{dv_{gfq}(t)}{dt} + \omega_o C_{gf}v_{gfd}(t) = i_{giq}(t) - i_{lq}(t) \\ C_{gf}\frac{dv_{gf0}(t)}{dt} = i_{gi0}(t) - i_{l0}(t) \end{cases}, \tag{3.16}$$

where the subscripts d and q refer to variables on the respective axes. As a result, all the signals are dc components in steady state. Notably, as all the coefficients are constants (e.g., V_{gdc_ref}), the UPS plant model described by (3.15) and (3.16) is linear in its large-signal form.

For balanced three-phase systems, the 0-axis equations can be ignored [i.e., ignorance of the third equations of (3.15) and (3.16)]. On top of that, we remove steady-state relationships and focus only the perturbations of (3.15) and (3.16), which are represented by the same equations with perturbed variables as

$$\begin{cases} V_{gdc\_ref}\Delta S_{gd}(t) = \Delta v_{gfd}(t) + L_{gi}\frac{d\Delta i_{gid}(t)}{dt} + R_{gi}\Delta i_{gid}(t) - \omega_o L_{gi}\Delta i_{giq}(t) \\ V_{gdc\_ref}\Delta S_{gq}(t) = \Delta v_{gfq}(t) + L_{gi}\frac{d\Delta i_{giq}(t)}{dt} + R_{gi}\Delta i_{giq}(t) + \omega_o L_{gi}\Delta i_{gid}(t) \end{cases} \text{and} \tag{3.17}$$

$$\begin{cases} C_{gf}\frac{d\Delta v_{gfd}(t)}{dt} - \omega_o C_{gf}\Delta v_{gfq}(t) = \Delta i_{gid}(t) - \Delta i_{ld}(t) \\ C_{gf}\frac{d\Delta v_{gfq}(t)}{dt} + \omega_o C_{gf}\Delta v_{gfd}(t) = \Delta i_{giq}(t) - \Delta i_{lq}(t) \end{cases}. \tag{3.18}$$

Finally, we derive the complex frequency domain expressions of the above model through the Laplace transform:

$$\begin{cases} \Delta i_{gid}(s) = \frac{1}{L_{gi}s+R_{gi}}\left[V_{gdc\_ref}\Delta S_{gd}(s) - \Delta v_{gfd}(s) + \omega_o L_{gi}\Delta i_{giq}(s)\right] \\ \Delta i_{giq}(s) = \frac{1}{L_{gi}s+R_{gi}}\left[V_{gdc\_ref}\Delta S_{gq}(s) - \Delta v_{gfq}(s) - \omega_o L_{gi}\Delta i_{gid}(s)\right] \end{cases} \text{and} \tag{3.19}$$

$$\begin{cases} \Delta v_{gfd}(s) = \frac{1}{C_{gf}s}\left[\Delta i_{gid}(s) - \Delta i_{ld}(s) + \omega_o C_{gf}\Delta v_{gfq}(s)\right] \\ \Delta v_{gfq}(s) = \frac{1}{C_{gf}s}\left[\Delta i_{giq}(s) - \Delta i_{lq}(s) - \omega_o C_{gf}\Delta v_{gfd}(s)\right] \end{cases}. \tag{3.20}$$

The above model of UPS system plants is shown in Fig. 3.12, where the cross-coupling terms (e.g., $-\omega_o L_{gi} i_{gid}$) are introduced by frame transformations. Such cross-coupling terms will affect system stability, as will be discussed in the next chapter. With cross-coupling terms ignored, we can design controllers independently in the d-axis and q-axis according to the classic control theory. By adjustment of switch functions, we can regulate ac voltages v_{gfabc}.

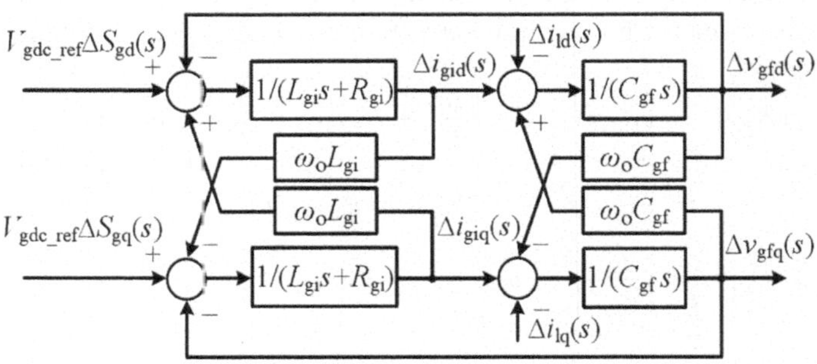

Fig. 3.12 Block diagram of UPS system plants

So far, we have detailed the modeling of UPS plants, which are typical three-phase LC-filtered system plants. Other three-phase dc-ac or ac-dc power converter plants can be modeled in a similar way, and hence will not be repeated. Notably, the aforementioned modeling procedure is essentially referred to state-space average modeling, which we have discussed in Chap. 1. The only difference is that the model has not been expressed in a state-space form for simplicity. In addition, we have not encountered any nonlinearity due to the assumption of constant dc buses. This assumption is not necessarily satisfied in other grid-tied power converters, which will be discussed later.

3.2.3 Controller Design

This subsection aims to design the voltage and current controllers of UPSs according to the system parameters listed in Table 3.2. In this subsection, the voltage and current controllers will be designed in the discrete z-domain to illustrate how to design digital controllers.

Figure 3.13 presents the block diagram of voltage and current controllers in the mixed s- and z-domains, where $G_{gv}(z)$ and K_{gf} denote the voltage regulator and voltage feedforward compensator, respectively. Specifically, we implement the voltage regulator as a PI + repetitive controller to reject harmonics. K_{gc} and K_{gl} represent the current regulator and current feedback compensator gains, respectively. z^{-1} and the ZOH (zero-order hold) model the PWM generator. To be specific, the

Table 3.2 UPS system parameters

Descriptions	Symbols	Values
Filter capacitance	C_{gf}	50 μF
Filter inductance	L_{gi}	1 mH
Dc-link voltage reference	V_{gdc_ref}	400 V
Ac voltage reference (rms)	V_{gf_ref}	110 V
Fundamental frequency	f_o	50 Hz
Switching/sampling frequency	f_{sw}/f_s	10 kHz

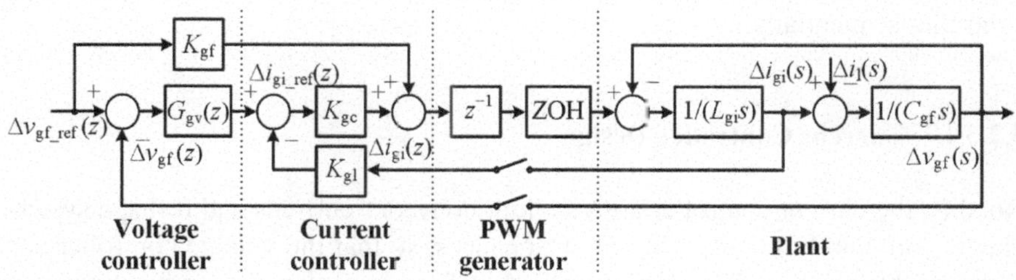

Fig. 3.13 Block diagram of UPS voltage and current controllers in the mixed s- and z-domains

one sampling period delay z^{-1} is caused by the calculation and update of reference signals [20]. After being updated, modulation waves should be compared with carrier waves. This process can approximately be modelled by a ZOH, which is equivalent to a 0.5 sampling-period delay [20]. However, it should be kept in mind that this approximation only holds valid in low-frequency regions, where frequencies are far below the switching and sampling frequencies. Referring to Table 3.2, the switching and sampling frequencies are identical in this case.

As observed from Fig. 3.13, the plant model is the simplified version of that in Fig. 3.12, where coupling effects and resistance R_{gi} are ignored. Along with coupling effects, the subscripts d and q are removed. Note that we can combine the constant V_{gdc_ref} in Fig. 3.12 in controllers.

The plant transfer functions from the output of the ZOH to Δi_{gi} and Δv_{gf} are respectively derived as

$$G_{inv\_i}(s) = \frac{\Delta i_{gi}(s)}{\Delta v_{gi}(s)} = \frac{C_{gf}s}{L_{gi}C_{gf}s^2 + 1} \text{ and} \tag{3.21}$$

$$G_{inv\_v}(s) = \frac{\Delta v_{gf}(s)}{\Delta v_{gi}(s)} = \frac{1}{L_{gi}C_{gf}s^2 + 1}, \tag{3.22}$$

Applying z transforms with the ZOH to (3.21) and (3.22), we obtain

$$G_{inv\_i}(z) = \frac{\Delta i_{gi}(z)}{\Delta v_{gi}(z)} = \frac{\sin(\omega_{gr}T_s)(z-1)}{L_{gi}\omega_{gr}\left[z^2 - 2\cos(\omega_{gr}T_s)z + 1\right]} \text{ and} \tag{3.23}$$

$$G_{inv\_v}(z) = \frac{\Delta v_{gf}(z)}{\Delta v_{gi}(z)} = \frac{\left[1 - \cos(\omega_{gr}T_s)\right](z+1)}{z^2 - 2\cos(\omega_{gr}T_s)z + 1}, \tag{3.24}$$

where T_s (i.e., $1/f_s$) stands for the sampling period. ω_{gr} denotes the resonance angular frequency of LC filters, given by

$$\omega_{gr} = \frac{1}{\sqrt{L_{gi}C_{gf}}}. \tag{3.25}$$

Using (3.23) and (3.24), we obtain the discrete z-domain block diagram, as depicted in Fig. 3.14. Next, we start to design the current controller and voltage controller sequentially.

3.2.3.1 Current Controller Design

Notably, the current controller aims to limit converter currents and reshape system plants, particularly for damping LC resonances, so that the voltage controller can better be designed. In this sense, we should take case of the transfer function from the current reference Δi_{gi_ref} to the capacitor voltage Δv_{gf}, which is derived from

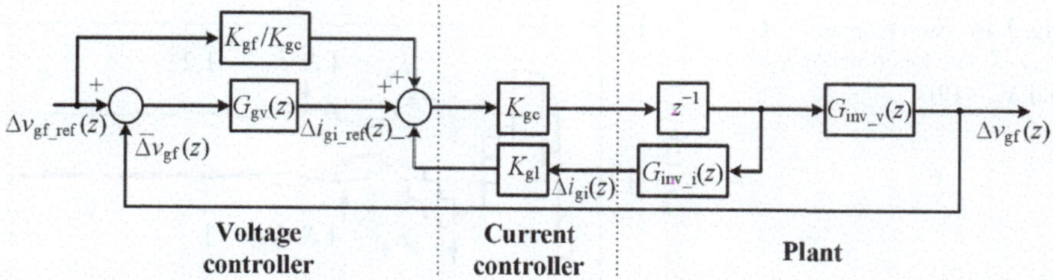

Fig. 3.14 Block diagram of UPS voltage and current controllers in the z-domain

Fig. 3.14 as

$$T_{iref\_v}(z) = \frac{\Delta v_{gf}(z)}{\Delta i_{gi\_ref}(z)} = \frac{K_{gc} G_{inv\_v}(z)}{z + K_{gl} K_{gc} G_{inv\_i}(z)}. \tag{3.26}$$

It is apparent from (3.26) that the product of K_{gl} and K_{gc} (i.e., $K_{gl}K_{gc}$) determines the denominator of $T_{iref\_v}(z)$, while K_{gc} affects the gain (or the numerator) of $T_{iref\_v}(z)$. To illustrate, Fig. 3.15 shows the pole-zero map of $T_{iref\_v}(z)$ [19]. As $(K_{gl}K_{gc})$ increases, $T_{iref\_v}(z)$ is gradually stabilized and then becomes unstable again, as one pair of its conjugate poles shifts gradually inside the unit circle and then outwards. As mentioned, stable discrete systems have all poles inside the unit circle.

To yield satisfactory stability margins and fast dynamics, we intend to design the conjugate poles as close as possible to the origin of the unit circle, thereby leading to the case of $(K_{gl}K_{gc}) = 4$. After determining $(K_{gl}K_{gc})$, we further design K_{gl} and K_{gc} individually. Notably, a steady-state gain of $T_{iref\_v}(z)$ as one will be helpful for voltage controller design. In steady state, substitution of $z = 1$ into (3.23), we know that the

Fig. 3.15 Pole-zero map of $T_{iref\_v}(z)$ as a function of $(K_{gl}K_{gc})$ [19]

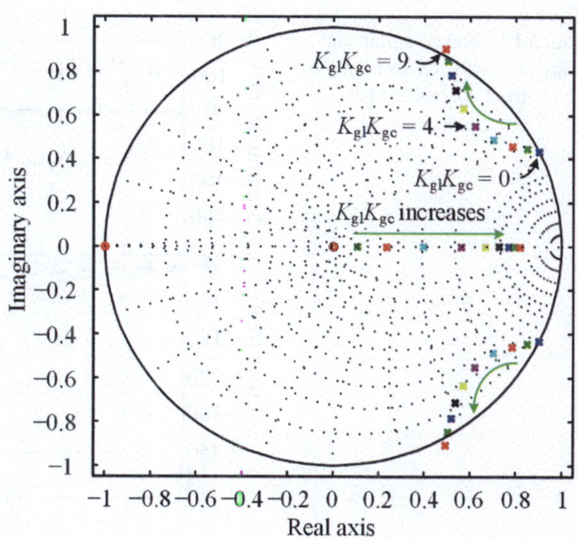

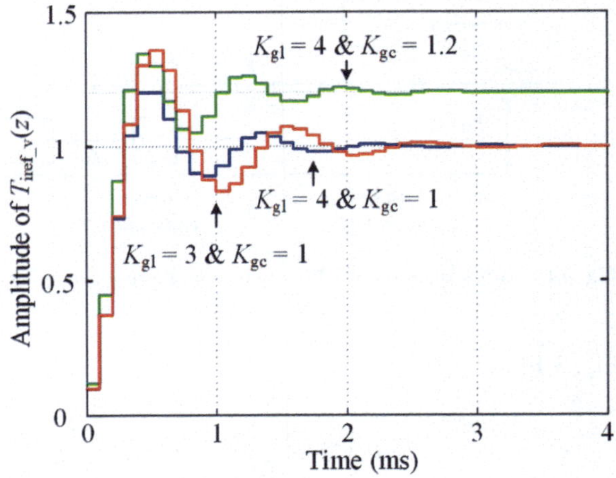

Fig. 3.16 Step responses of $T_{iref\_v}(z)$ as functions of K_{gl} and K_{gc} [19]

steady-state value of $G_{inv\_i}(z)$ is 0. By substituting $G_{inv\_i}(z) = 0$ and $z = 1$ into (3.26), we find that the steady-state value of $T_{iref\_v}(z)$ is independent of K_{gl}. Therefore, we can design a unity steady-state gain by choosing K_{gc}. Figure 3.16 shows the step responses of $T_{iref\_v}(z)$ as functions of K_{gl} and K_{gc} [19]. Clearly, the steady-state gain of $T_{iref\_v}(z)$ depends only on K_{gc}. By designing $K_{gc} = 1$, the steady-state gain of $T_{iref\_v}(z)$ is unity.

Figure 3.17 demonstrates the Bode diagrams of $T_{iref\_v}(z)$, showing the effect current controller on reshaping the system plant. Without the current controller (i.e., $K_{gl} = 0$ and $K_{gc} = 1$), we notice an obvious resonant peak introduced by the LC filter. With the designed current controller (i.e., $K_{gl} = 4$ and $K_{gc} = 1$), the resonant peak is successfully damped. Moreover, $T_{iref\_v}(z)$ features a magnitude of 0 dB in

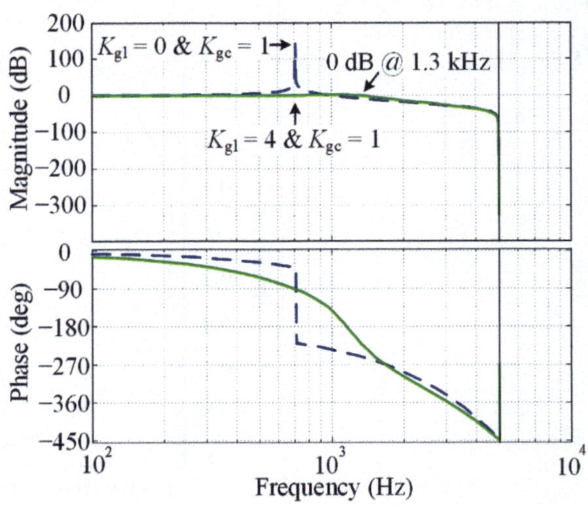

Fig. 3.17 Bode diagrams of $T_{iref\_v}(z)$ with and without the current controller [19]

the low-frequency band until the frequency reaches 1.3 kHz. After being designed, the current controller and plant are replaced and represented by $T_{iref\_v}(z)$.

3.2.3.2 Voltage Controller Design

The voltage controller aims to eliminate voltage tracking errors and reject disturbances. Considering that the voltage reference v_{gf_ref} is a constant V_{gf_ref} in the d-axis and zero in the q-axis, a PI controller is sufficient for voltage tracking without any error according to the IMP, as detailed in Chap. 1. On top of that, we incorporate a paralleled repetitive controller to reject harmonic disturbances, which normally come from nonlinear loads [21].

Figure 3.18 presents the voltage controller, where the voltage regulator $G_{gv}(z)$ is further detailed as a combination of a PI controller $G_{gv\_pi}(z)$ and a repetitive controller, which comprises a fundamental period delay unit $z^{-N_{gr}}$, a constant around one Q_{gr}, a low-pass filter $S_{gr}(z)$, a phase compensation unit $z^{N_{gc}}$, and a proportional gain K_{gr} [22]. Clearly, this practical repetitive controller is different from the ideal one in (1.28). First, the repetitive controller is implemented in the discrete z domain, where the fundamental period continuous delay e^{-sT0} is replaced by the discrete delay $z^{-N_{gr}}$. Second, the stabilized factor Q_{gr} is used instead of one to avoid closed-loop poles being located on the unity circle. Third, the low-pass filter $S_{gr}(z)$ and its phase compensation unit $z^{N_{gc}}$ and gain compensation unit K_{gr} act together to reject high-frequency noise while maintaining a zero phase shift.

In Fig. 3.18, the voltage feedforward compensator K_{gf} remains unchanged. We have designed $T_{iref\_v}(z)$ so that it exhibits a unity gain in the low-frequency band. Therefore, K_{gf} is designed to be one so that the feedforward voltage gain is unity. Notably, this feedforward voltage control facilitates voltage tracking and disturbance rejection.

First, we design the repetitive controller for harmonic rejection and ignore the PI controller. To obtain a fundamental period time-delay, $N_{gr} = f_s/f_{ref} = 200$ is derived. Next, we express the voltage error Δv_{gf_err} as a function of the voltage reference Δv_{gf_ref} as

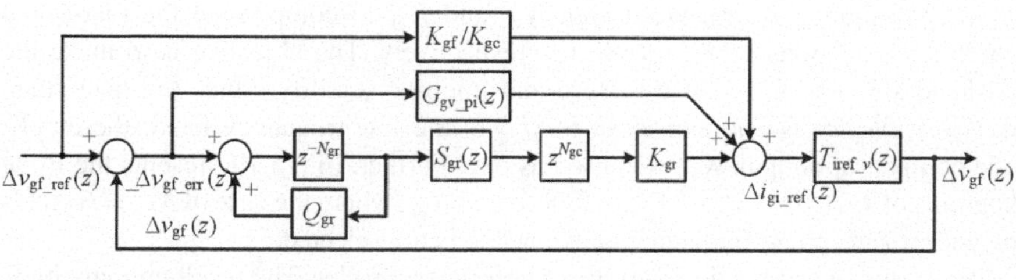

Fig. 3.18 Detailed block diagram of the UPS voltage controller

$$\Delta v_{\text{gf\_err}}(z) = \frac{\Delta v_{\text{gf\_ref}}(z)(z^{N_{\text{gr}}} - Q_{\text{gr}})\left[1 - T_{\text{iref\_v}}(z)\right]}{(z^{N_{\text{gr}}} - Q_{\text{gr}}) + S_{\text{gr}}(z)z^{N_{\text{gc}}}K_{\text{gr}}T_{\text{iref\_v}}(z)}, \tag{3.27}$$

whose characteristic equation can be obtained as

$$z^{N_{\text{gr}}} - Q_{\text{gr}} + S_{\text{gr}}(z)z^{N_{\text{gc}}}K_{\text{gr}}T_{\text{iref\_v}}(z) = 0. \tag{3.28}$$

To ensure stable UPS systems, all the roots of (3.18) should be inside the unity circle. Under this condition, $|z| < 1$ should be satisfied, leading to

$$\left|S_{\text{gr}}(z)z^{N_{\text{gc}}}K_{\text{gr}}T_{\text{iref\_v}}(z) - Q_{\text{gr}}\right| < 1, \tag{3.29}$$

which is a sufficient condition for system stability.

Next, we derive harmonic voltage errors. By inserting $z = e^{j(2k\pi/N\text{gr})}$ $(k = 1, 2, 3, \ldots, N_{\text{gr}}/2)$ into (3.27), we obtain

$$\left|\Delta v_{\text{gf\_err}}(e^{j2k\pi/N_{gr}})\right| = \frac{\Delta v_{\text{gf\_ref}}(e^{j2k\pi/N_{\text{gr}}})(1 - Q_{\text{gr}})\left[1 - T_{\text{iref\_v}}(e^{j2k\pi/N_{\text{gr}}})\right]}{1 - \left[Q_{\text{gr}} - S_{\text{gr}}(e^{j2k\pi/N_{\text{gr}}})e^{j2k\pi N_{\text{gc}}/N_{\text{gr}}}K_{\text{gr}}T_{\text{iref\_v}}(e^{j2k\pi/N_{\text{gr}}})\right]}, \tag{3.30}$$

where k denotes the harmonic order. We know from (3.30) that there are two obtains to achieve zero harmonic errors. The first option is $Q_{\text{gr}} = 1$, and this is the case of ideal repetitive controllers. To yield stable systems, we select Q_{gr} as a constant around one (e.g., 0.95), thus failing to completely suppress harmonics. The second option refers to $T_{\text{iref\_v}} = 1$. This condition can hold valid in the low-frequency band through a well-designed current controller, as detailed and shown in Fig. 3.18. Another remainder is that the denominator of (3.30) should stay away from zero to avoid harmonic amplifications.

As mentioned, the low-pass filter $S_{\text{gr}}(z)$ aims to suppress high-frequency noise. We design the cut-off frequency of $S_{\text{gr}}(z)$ as 500 Hz, considering the trade-off between control bandwidth and harmonic rejection. A second-order Butterworth low-pass filter is a common option. However, other implementations of $S_{\text{gr}}(z)$, such as notch filters, are also possible [21].

After designing $S_{\text{gr}}(z)$, we design N_{gc} and K_{gr} to compensate the phase and amplitude deviations of $T_{\text{iref\_v}}(z)S_{\text{gr}}(z)$, respectively. The objective is to make the left-hand side of (3.29) close to zero and improve stability. Since the magnitude of $T_{\text{iref\_v}}(z)S_{\text{gr}}(z)$ is one and close to Q_{gr} in the low-frequency band, the amplitude compensation gain K_{gr} is chosen as one. Further, Fig. 3.19 presents the Bode diagrams of $T_{\text{iref\_v}}(z)S_{\text{gr}}(z)z^{N\text{gc}}$ as a function of N_{gc}, where the case of $N_{\text{gc}} = 6$ yields the widest zero-phase frequency range, hence being selected.

We have finished the design of the repetitive controller. The resultant magnitude plot of the system loop gain $\Delta v_{\text{gf}}/\Delta v_{\text{gf\_err}}$ is depicted in Fig. 3.20, where the peaks at harmonic frequencies are observed as expected.

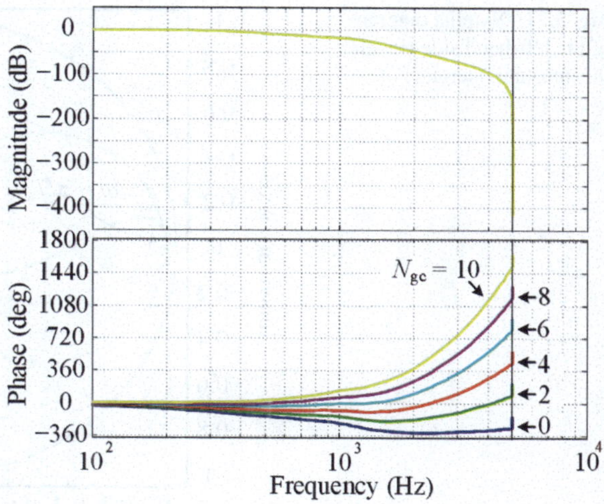

Fig. 3.19 Bode diagrams of $T_{iref\_v}(z)S_{gr}(z)z^{N_{gc}}$ as a function of N_{gc} [19]

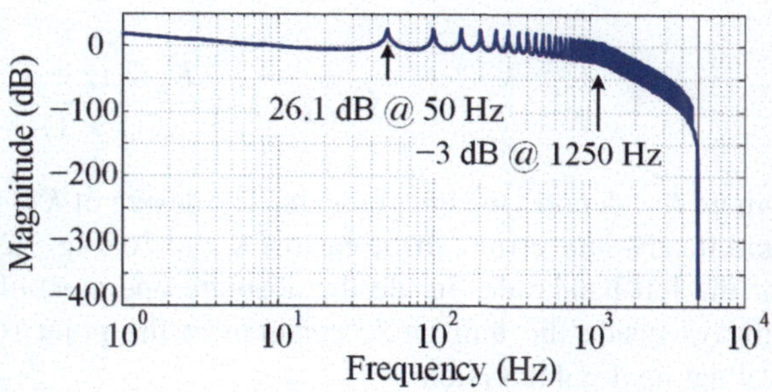

Fig. 3.20 Bode diagram of $\Delta v_{gf}(z)/\Delta v_{gf\_err}(z)$ [19]

In what follows, we analyze the influence of the additional PI controller on system stability. Taking the PI controller into consideration, we rewrite the voltage error expression as

$$\Delta v_{gf\_err}(z) = \frac{\Delta v_{gf\_ref}(z)(z^{N_{gr}} - Q_{gr})\left[1 - T_{iref\_v}(z)\right]}{(z^{N_{gr}} - Q_{gr})\left[1 + T_{iref\_v}(z)G_{gv\_pi}(z)\right] + S_{gr}(z)z^{N_{gc}}K_{gr}T_{iref\_v}(z)}. \tag{3.31}$$

The sufficient condition for stable systems changes into

$$\left|\frac{S_{gr}(z)z^{N_{gc}}K_{gr}T_{iref\_v}(z)}{1 + T_{iref\_v}(z)G_{gv\_pi}(z)} - Q_{gr}\right| < 1. \tag{3.32}$$

As we can modify the repetitive controller into a P controller, only the case where $G_{gv\_pi}(z)$ is an I controller is considered. Through the bilinear z transformation, $G_{gv\_pi}(z)$ is discretized as

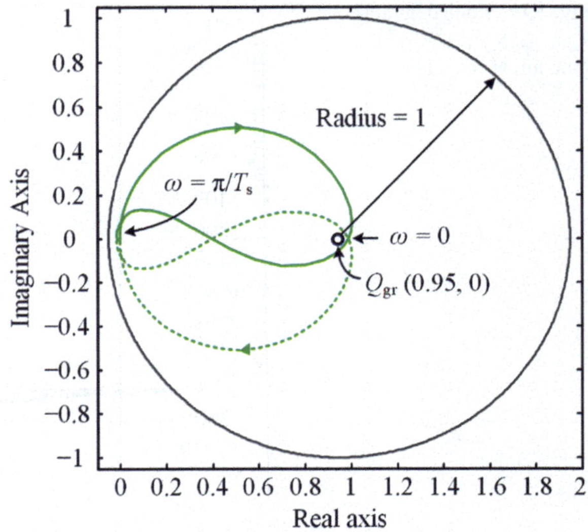

Fig. 3.21 Nyquist diagram of the left-hand side (inside the absolute operator) of (3.22) [19]

$$G_{gv\_pi}(z) = \frac{K_{gv} T_s (z+1)}{2(z-1)}, \tag{3.33}$$

where K_{gv} denotes the integral gain. The design of K_{gv} is a tradeoff between stability and steady-state errors. In the case of $K_{gv} = 70$, Fig. 3.21 shows the Nyquist diagram of the left-hand side (inside the absolute operator) of (3.22). Clearly, the trace is always inside the unit circle centered at the point (Q_{gr}, 0) or (0.95, 0), thereby validating a stable system.

3.2.3.3 Results

Table 3.3 lists the control parameters of UPSs designed in the previous part. We

Table 3.3 UPS control parameters

Descriptions	Symbols	Values
Current feedback gain	K_{gl}	4
Current regulator gain	K_{gc}	1
Voltage feedforward gain	K_{gf}	1
Fundamental period delay number	N_{gr}	200
Stabilized factor	Q_{gr}	0.95
Cut-off frequency of low-pass filter	f_{gr_cut}	500 Hz
Phase compensation number	N_{gc}	6
Magnitude compensation gain	K_{gr}	1
Voltage integral gain	K_{gv}	70

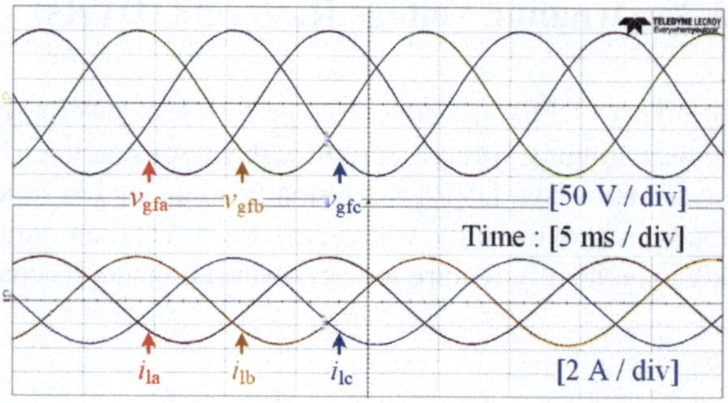

Fig. 3.22 Experimental waveforms of a UPS supplying a resistor load [19]

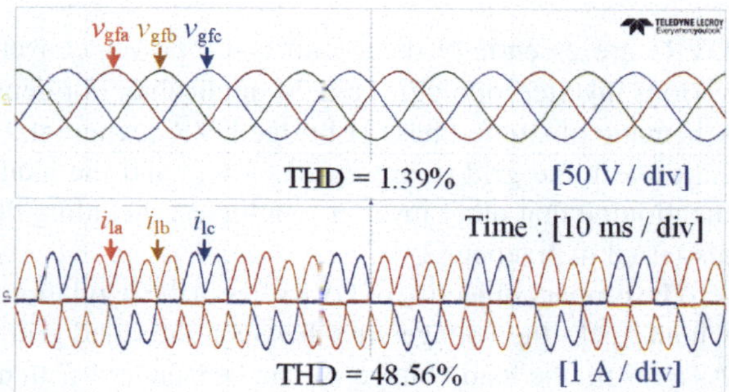

Fig. 3.23 Experimental waveforms of a UPS supplying a nonlinear load [19]

conducted experiments on UPSs, which were designed based on the parameters given in Tables 3.2 and 3.3.

Figure 3.22 shows the experimental results of a UPS, which supplies a linear resistor load. As expected, the capacitor voltage waveforms v_{gfabc} are highly sinusoidal and symmetrical.

Figure 3.23 provides the experimental results of a UPS fed by a nonlinear diode-rectifier load, as shown in the right-hand part of Fig. 2.3. Once again, the capacitor voltages are highly sinusoidal and symmetrical even in the face of a nonlinear load. In this case, the nonlinear load features a current THD of 48.56%. With properly designed controllers, the UPS manages to maintain the voltage THD as low as 1.39%, thereby justifying the effectiveness of the proposed voltage and current controllers. To analyze THDs, voltage/current waveforms were exported from an oscilloscope and then analyzed by the Matlab/Simulink Powergui FFT Analyze Tool. These experimental results clearly demonstrate the feasibility and effectiveness of UPSs.

3.3 Dynamic Voltage Restorers (DVRs)

Similar to UPSs, dynamic voltage restorers (DVRs) target at voltage quality problems, including voltage amplitude deviations, voltage harmonics, and voltage imbalances. However, DVRs are normally connected in series (instead of in parallel) with loads to provide only voltage errors rather than entire load voltages. As a result, DVRs generally feature lower power ratings as compared to UPSs.

3.3.1 Fundamental Principles

DVRs are essentially dc-ac and ac-dc power converters. In terms of topologies, various inverter topologies can be applicable. For demonstration, Fig. 3.24 shows a schematic of single-phase H-bridge DVRs, where the filter capacitor of DVRs is in series with the grid (represented by v_{gs}) and the load (denoted by v_{gl}). It is worth mentioning that other inverter topologies, including the ones with transformers, are also used in practice [23].

The primary objective of DVRs lies in the regulation of the ac capacitor voltage v_{gf}, which tracks the voltage error between the stand grid voltage and real grid voltage. As a result, the load voltage v_{gl} can be standard with high quality.

To ensure a high-quality load voltage, DVRs should accurately detect voltage errors from either the grid voltage or the load voltage. In this sense, the voltage error measurement unit is an important part in DVRs, which gives the reference for capacitor voltage controllers. As will be introduced, similar measurement units are employed in APFs. Ideally, voltage and current measurements should be fast and accurate. Next, we discuss the implementations of voltage error measurements.

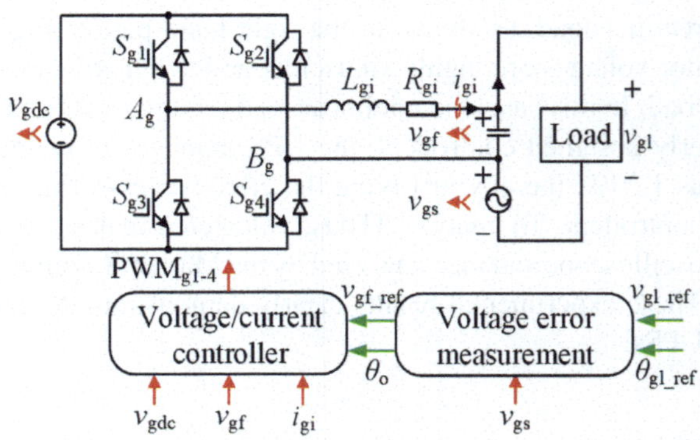

Fig. 3.24 Schematic of single-phase H-bridge DVRs

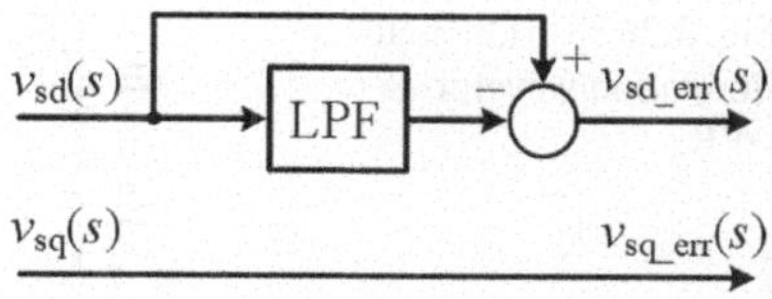

Fig. 3.25 Block diagram of *dq*0-frame voltage error measurement units

3.3.2 Voltage Error Measurement Units

We explore voltage error measurement units in this subsection. For three-phase systems, the detection of voltage errors through frame rotations is popular. After *abc*-*dq*0 transformations, instantaneous voltage signals are mapped to their *dq*0 components. As mentioned, a constant component in the *d*-axis represents the fundamental positive-sequence voltage component. Therefore, we use low-pass filters to extract the constant *d*-axis component, which is subsequently being subtracted from *dq*0 voltage components. The resultant signals are then inversely transformed into the *abc*-frame and form instantaneous voltage errors. Figure 3.25 illustrates the block diagram of *dq*0-frame harmonic measurement units.

For single-phase systems, we can construct virtual three-phase signals by delaying single voltage signals (e.g., by 120° and 240°). Subsequently, voltage errors are extracted as in Fig. 3.25. In addition, second-order generalized integrators (SOGIs) may also generate *dq*-frame voltage signals [24]. In addition to the *dq*-frame detection method, FFT, wavelet transform, and neural network-based detection methods are other options. After measurements, voltage errors become the references of voltage controllers, which regulate ac capacitor voltages to follow voltage errors so that load voltages become highly sinusoidal and symmetrical with nominal amplitudes.

3.3.3 Results

As for control, DVRs and UPSs share similar controllers and structures, and hence not repeated here. However, in DVRs, voltage controllers may possibly track harmonics instead of rejecting them as in UPSs. Moreover, grid voltages are often regarded as disturbances in DVRs.

With properly designed controllers, Figs. 3.26 and 3.27 demonstrates the steady-state and dynamic simulation results of DVRs, respectively. Figure 3.26 illustrates a case where the grid voltage contains 20% third-order and 8% fifth-order harmonics, leading to a THD of 22.01%. After compensation, the load voltage is standard with a THD of 1.95%, thereby demonstrating the effectiveness of voltage compensation.

Figure 3.27 shows dynamic simulation results of DVRs, where a grid voltage swell of 40 V occurs at 0.1 s, followed by a voltage sag 40 V at 0.4 s, and then restores at 0.7 s. Regardless of grid voltages, the load voltage remains as a standard voltage, indicating the effectiveness of DVRs.

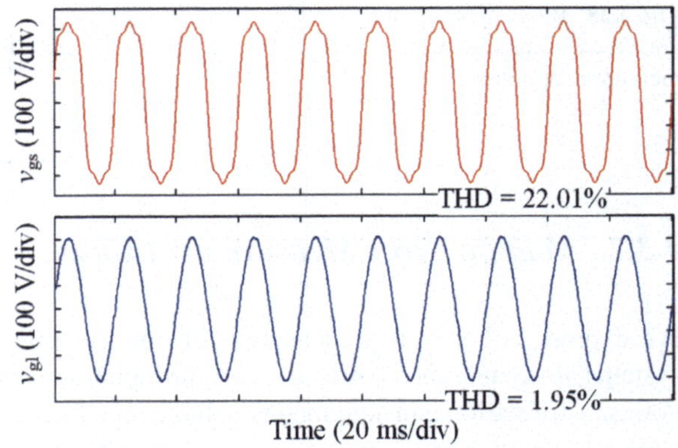

Fig. 3.26 Steady-state simulation waveforms of DVRs

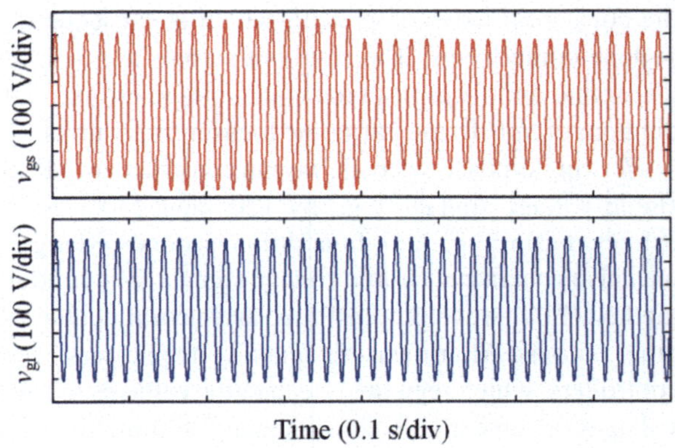

Fig. 3.27 Dynamic simulation waveforms of DVRs

As introduced, DVRs are effective voltage quality conditioning equipment. Alternative names of DVRs include series APFs and active voltage quality regulators (AVQRs). Next, we will investigate current quality enhancement equipment.

3.4 Static Compensators (STATCOMs)

Static compensators (STATCOMs) or distributed static compensators (DSTATCOMs) are the ac-dc and dc-ac grid-tied power converters that aim to compensate reactive currents (or reactive power). As voltage amplitude deviations depend on reactive power balance [1], reactive power compensation also contribute to the improvement of voltage quality. STATCOMs are sometimes also called static var generators (SVGs) [25].

3.4.1 Fundamental Principles

Generally, STATCOMs are connected in parallel to loads and grids through PCCs. They compensate reactive power according to load currents or reactive power references. However, STATCOMs, particularly for DSTATCOMs, are increasingly implemented by existing inverters. For demonstration, Fig. 3.28 shows a schematic diagram of three-phase STATCOMs, where the dc bus is represented as a constant voltage source. Notably, it should be replaced by a capacitor if without any external regulation, which will be discussed later.

In Fig. 3.28, the power grid is modelled as an ideal three-phase voltage source, whose voltages are denoted as v_{gabc}. In this case, we ignore the line impedance for simplicity. As mentioned, we use either the *L* filter or the *LCL* filter as a grid interface. The primary objective of STATCOMs is to regulate the grid currents i_{cgabc} as per their references i_{cgd_ref} and i_{cgq_ref}.

Figure 3.29 shows a phasor diagram of STATCOMs. Noticeably, STATCOMs compensate the reactive current component in the load via $\mathbf{i}_{\mathbf{statcom}}$, resulting in the in phase grid voltage and current phasors $\mathbf{v}_{\mathbf{gird}}$ and $\mathbf{i}_{\mathbf{grid}}$.

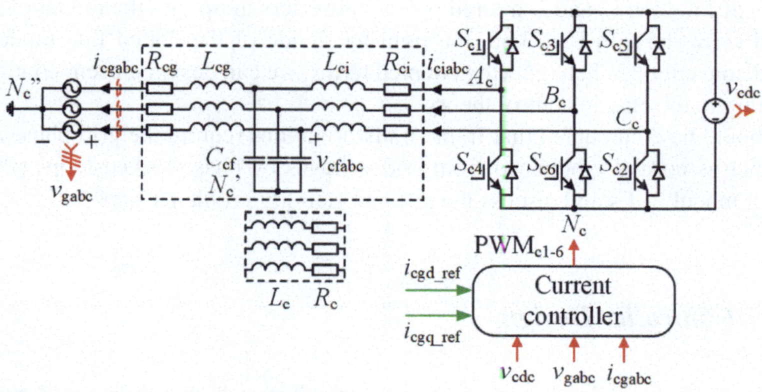

Fig. 3.28 Schematic of three-phase STATCOMs

Fig. 3.29 Phasor diagram of STATCOMs

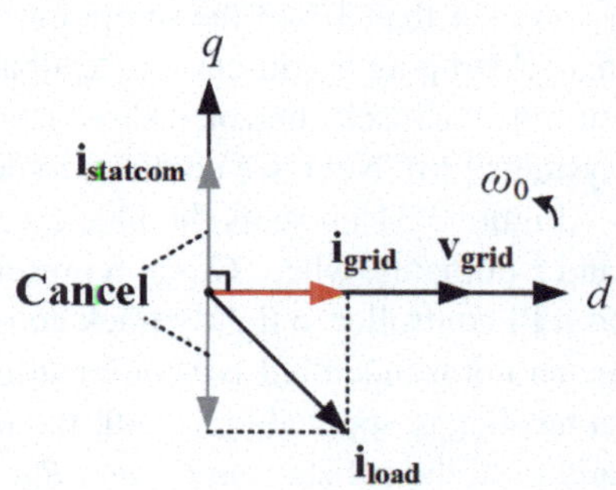

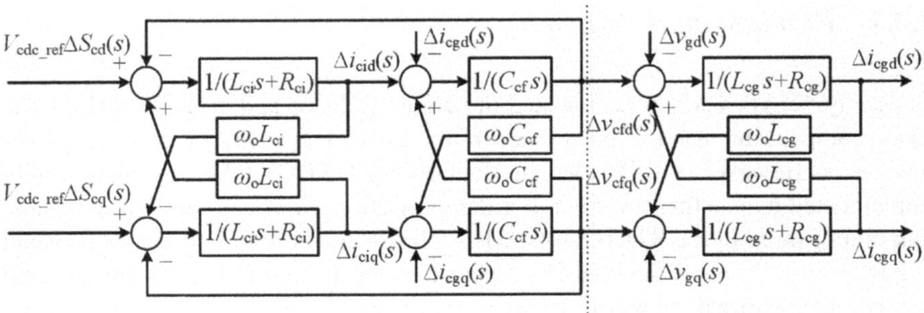

Fig. 3.30 Block diagram of STATCOM system plants

3.4.2 Plant Modeling

Following the modeling procedure in Sect. 3.4, we derive the plant model of STATCOMs, as shown in Fig. 3.30, where ΔS_{cd} and ΔS_{cq} represent switch functions in the *d*- and *q*-axes, respectively. V_{cdc_ref} refers to the rated dc voltage. The models of *LC*-filtered and *L*-filtered power converters lump into the model of *LCL*-filtered converters. In Fig. 3.30, the right-hand side of the dotted line models *L*-filtered converters. When ignoring coupled terms, we can design current controllers according to the classic control theory.

It should be commented that frame transformations require the grid phase angle θ_o, which is normally obtained from grid voltages by phase-locked-loops (PLLs). We will model PLLs and discuss the relevant stability problems later.

3.4.3 Controller Design

We mainly focus on the design of current controllers, which regulate grid currents to follow their references. We design controllers based on the system parameters in Table 3.4, where the parameters of the *LCL* filter yields a resonance frequency f_{cr} greater than 1/6 of the sampling frequency f_s. In this case, we can remove resonance damping if grid-current-feedback control is used [26]. In the case of converter-current-feedback control, it should be highlighted that $f_{cr} < f_s/6$ is required for stable systems [26]. Next, we continue to design discrete current controllers.

Figure 3.31 presents the block diagram of the current controller in the mixed *s*- and *z*-domains, where $G_{cc}(z)$ represents the current regulator. We implement $G_{cc}(z)$ as a PI controller to tightly track constant current references. Once again, the PWM generator is modelled as a combination of z^{-1} and the ZOH. Moreover, as the current controller is applicable to both the *d*- and *q*-axes, we ignore the *dq* subscripts. In addition, the resistances R_{ci} and R_{cg} are case dependent and often of small values, and hence ignored here.

Table 3.4 STATCOM system parameters

Descriptions	Symbols	Values
Filter capacitance	C_{cf}	5 μF
Grid filter inductance	L_{cg}	1 mH
Converter filter inductance	L_{ci}	1 mH
Total inductance	L_c	2 mH
DC-link voltage reference	V_{cdc_ref}	360 V
Grid voltage reference (rms)	V_{g_ref}	110 V
Fundamental frequency	f_o	50 Hz
Resonance frequency	f_{cr}	3.18 kHz
Switching/sampling frequency	f_{sw}/f_s	10 kHz

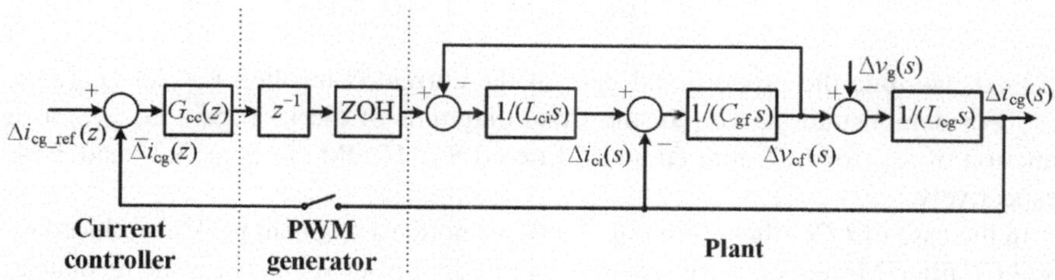

Fig. 3.31 Block diagram of STATCOM current controllers in the mixed *s*- and *z*-domains

We derive the transfer function from the output of the ZOH to the grid-injected current Δi_{cg} from Fig. 3.31 as

$$G_{plant}(s) = \begin{cases} \frac{\omega_{cr}^2}{(L_{ci}+L_{cg})s(s^2+\omega_{cr}^2)} & : LCL \\ 1/(L_c s) & : L \end{cases}, \tag{3.34}$$

where ω_{cr} denotes the resonance angular frequency of the *LCL* filter, i.e.,

$$\omega_{cr} = \sqrt{\frac{L_{ci}+L_{cg}}{L_{ci}L_{cg}C_{cf}}}. \tag{3.35}$$

Next, we derive the discrete plant transfer functions of (3.34) as

$$G_{plant}(z) = \begin{cases} \frac{T_s}{(L_{ci}+L_{cg})(z-1)} - \frac{\sin(\omega_{cr}T_s)(z-1)/(L_{ci}+L_{cg})}{\omega_{cr}\left[z^2-2\cos(\omega_{cr}T_s)z+1\right]} & : LCL \\ \frac{T_s}{L_c(z-1)} & : L \end{cases}. \tag{3.36}$$

Upon having $G_{plant}(z)$, we reorganize the block diagram of STATCOM current controllers in the *z* domain, as shown in Fig. 3.32.

We obtain the loop gain $G_{c\_ol}(z)$ from Fig. 3.32 as

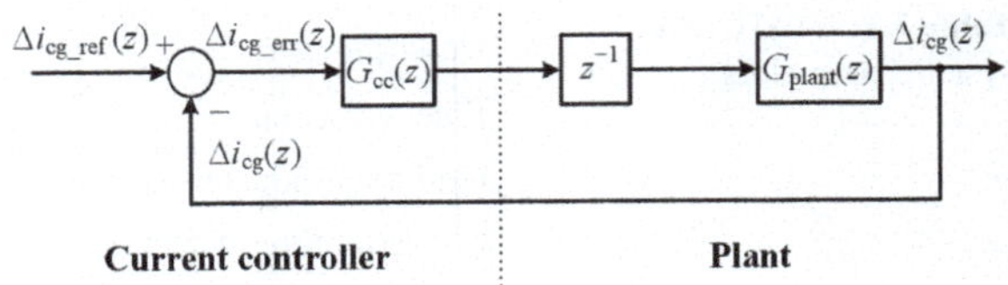

Fig. 3.32 Block diagram of STATCOM current controllers in the z-domain

$$G_{\mathrm{c\_ol}}(z) = \frac{\Delta i_{\mathrm{cg}}(z)}{\Delta i_{\mathrm{cg\_err}}(z)} = G_{\mathrm{cc}}(z) z^{-1} G_{\mathrm{plant}}(z), \tag{3.37}$$

based on which the closed-loop transfer function is expressed as

$$G_{\mathrm{c\_cl}}(z) = \frac{\Delta i_{\mathrm{cg}}(z)}{\Delta i_{\mathrm{cg\_ref}}(z)} = \frac{G_{\mathrm{c\_ol}}(z)}{1 + G_{\mathrm{c\_ol}}(z)}. \tag{3.38}$$

First, we tune the proportional gain of the current controller, i.e., let $G_{\mathrm{cc}}(z) = K_{\mathrm{cp}}$. For demonstration, we plot the Bode diagrams of the loop gain $G_{\mathrm{c\_ol}}(z)$ as a function of K_{cp} for *LCL*-filtered and *L*-filtered STATCOMs in Figs. 3.33 and 3.34, respectively.

In the case of *LCL* filters (see Fig. 3.33), we notice a resonant peak introduced by the *LCL* filter. Moreover, there are two magnitude zero-crossings near the resonance frequency. Fortunately, the phases at the zero-crossing frequencies are between $-180°$ and $-540°$, resulting in positive phase margins. Therefore, we evaluate the current controller by the GM and PM of the first zero-crossing frequency. Clearly, stability margins (i.e., the GM and PM) reduce as the proportional gain K_{cp} increases. To guarantee stability, GM > 3 dB and PM > 40° are desirable. In addition, a larger K_{cp} translates into faster dynamics. As a tradeoff, we choose the case $K_{\mathrm{cp}} = 10$, which gives a GM of 4.1 dB and PM of 44.4°.

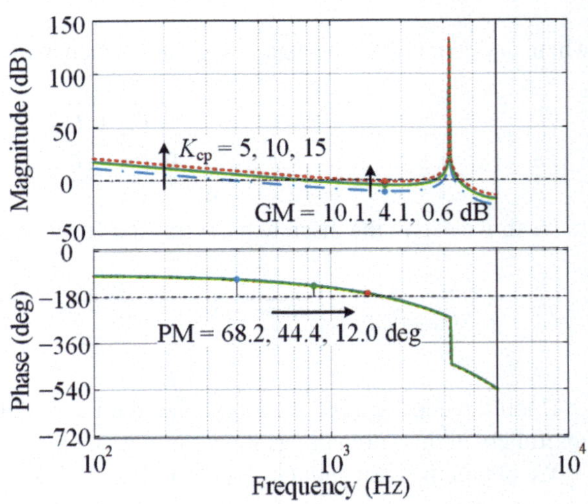

Fig. 3.33 Bode Diagram of $G_{\mathrm{c\_ol}}(z)$ as a function of K_{cp} for *LCL*-filtered STATCOMs

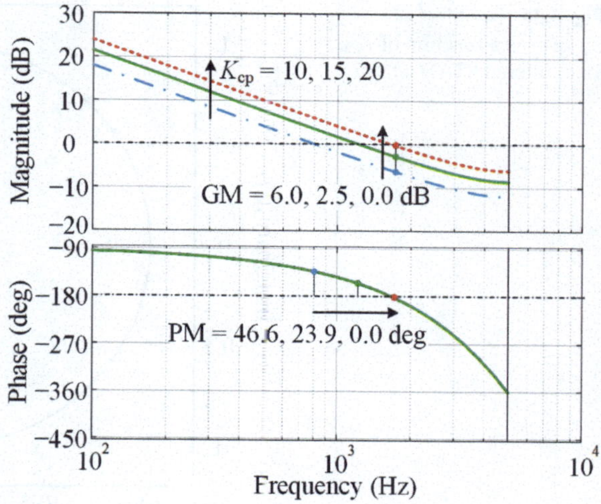

Fig. 3.34 Bode Diagram of $G_{c\_ol}(z)$ as a function of K_{cp} for L-filtered STATCOMs

In the case of L filters (see Fig. 3.34), there is no resonant peak. Correspondingly, stability margins are enlarged. Once again, the GM and PM decrease as K_{cp} increases. With a K_{cp} of 20, the system becomes critically stable, leading to instability in practice. Note that increasing L_c and decreasing K_{cp} have the same effect on the loop gain $G_{c\_ol}(z)$. Therefore, we can use a larger K_{cp} as L_c increases. For example, the case of $K_{cp} = 25$ leads to a stable system when $L_c = 5$ mH. Finally, K_{cp} is designed to be 15 for L-filtered STATCOMs.

We draw the root loci of $G_{c\_ol}(z)$ as a function of K_{cp} for LCL-filtered and L-filtered STATCOMs in Figs. 3.35 and 3.36, respectively. Notably, the maximum

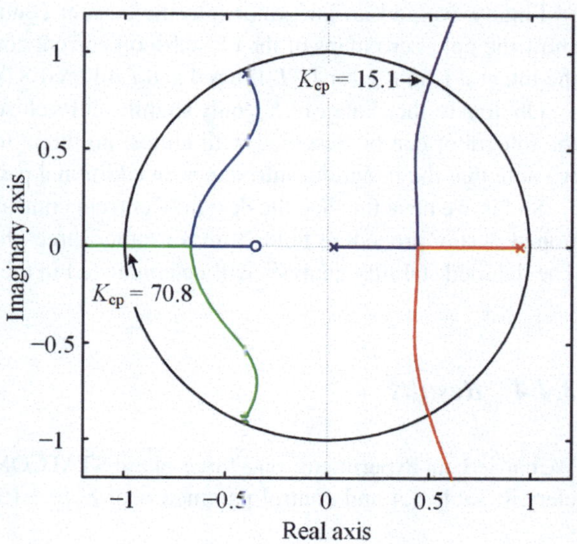

Fig. 3.35 Root loci of $G_{c\_ol}(z)$ as a function of K_{cp} for LCL-filtered STATCOMs

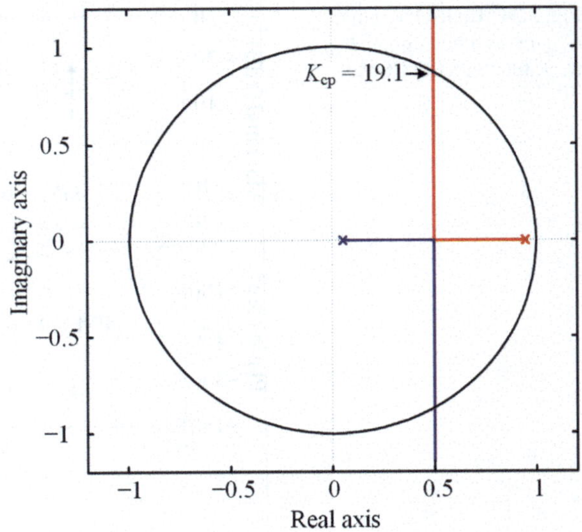

Fig. 3.36 Root loci of $G_{c\_ol}(z)$ as a function of K_{cp} for L-filtered STATCOMs

acceptable values of the proportional gain K_{cp} for stable current control can be identified. Specifically, $K_{cp} = 15.1$ and $K_{cp} = 19.1$ are the boundary conditions for LCL-filtered and L-filtered STATCOMs, respectively. Moreover, we observe that there are four poles and two poles in the cases of LCL filters and L filters, respectively. This can be understandable, as the time delay unit introduces one pole, while the remaining poles are caused by system plants. Additionally, it should be recognized that both root loci and pole-zero maps can visualize the locations of system poles. In comparison, root loci target at system loop gains, while pole-zero maps focus on closed-loop transfer functions.

Finally, we add an integrator into the current controller. Figures 3.37 and 3.38 show the pole-zero maps of the closed-loop current control $G_{c\_cl}(z)$ as a function of the integral gain K_{ci} for LCL-filtered and L-filtered STATCOMs, respectively.

Obviously, the change of K_{ci} only slightly shifts closed-loop poles. In other words, the integrator can be designed with almost no threat to system stability. Moreover, we note that the integrator introduces an additional pair of zero and pole.

So far, we have finished the design of current controllers. Note that the aforementioned design procedure holds true for either three-phase or single-phase systems. The detailed stability analysis will be conducted in the following chapters.

3.4.4 Results

We carried out experiments on a three-phase STATCOM based on the system parameters in Table 3.4 and control parameters of $K_{cp} = 15$ and $K_{ci} = 300$. Figure 3.39

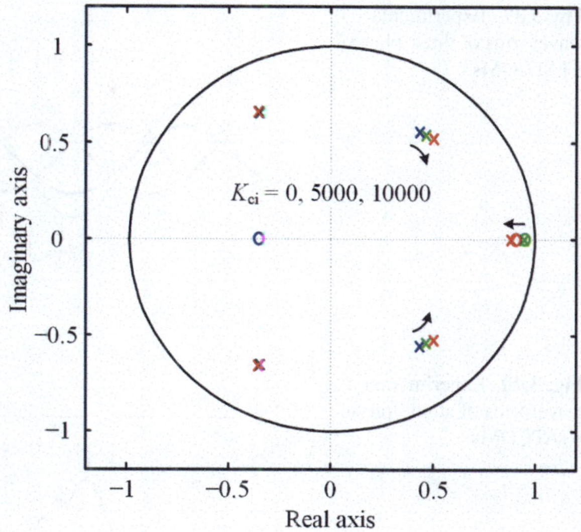

Fig. 3.37 Pole-zero map of $G_{c\_cl}(z)$ as a function of K_{ci} for *LCL*-filtered STATCOMs

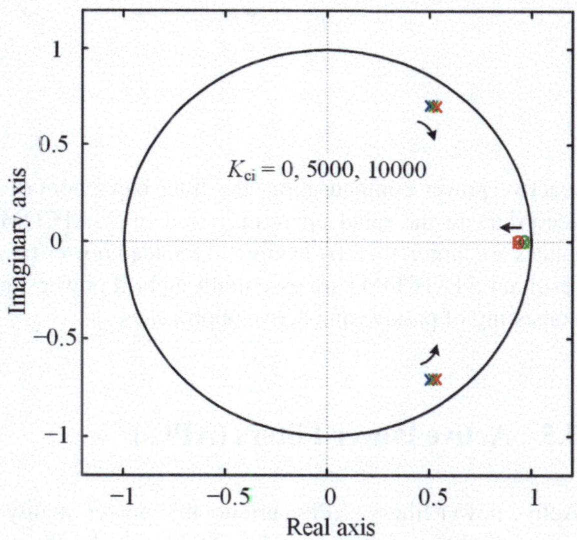

Fig. 3.38 Pole-zero map of $G_{c\_cl}(z)$ as a function of K_{ci} for *L*-filtered STATCOMs

shows the experimental results of current control, where the grid currents i_{cgabc} are sinusoidal and in phase with the grid voltages v_{gabc}.

Further, Fig. 3.40 demonstrates the experimental results of single-phase STATCOMs. Clearly, the STATCOM compensates the reactive current component in the load current i_{load}, leading to a unity power factor of the resultant grid current i_{grid}. These experiments validate the feasibility and effectiveness of STATCOMs.

It is worthwhile to note that the design of *LCL* filters in STATCOMs can be different from typical applications. As STATCOMs normally target at inductive

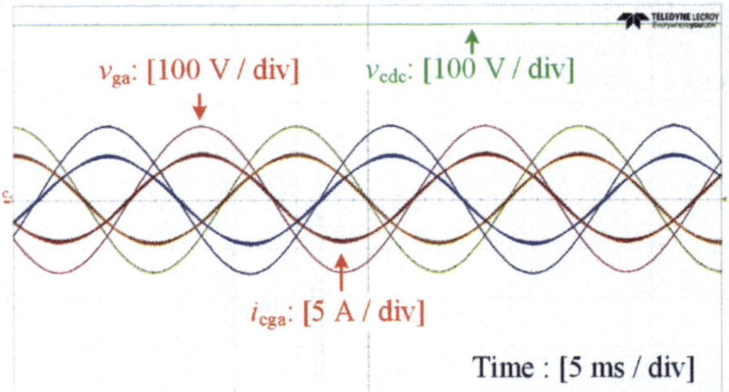

Fig. 3.39 Experimental waveforms of three-phase STATCOMs

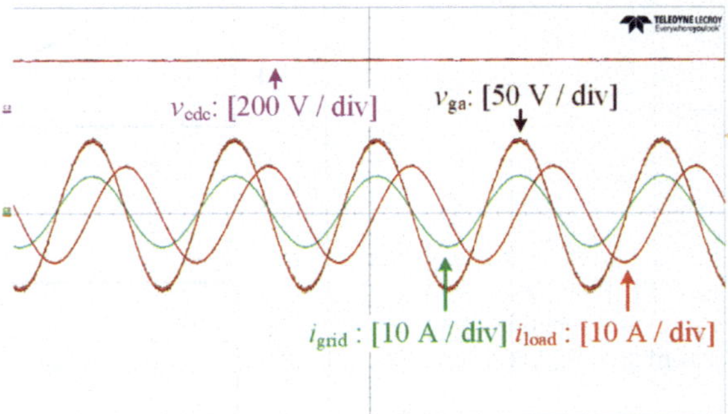

Fig. 3.40 Experimental waveforms of single-phase STATCOMs

reactive power compensation, the filter capacitors of *LCL* filters can be designed according to the rated operating point of STATCOMs. As a result, filter capacitances are larger, thereby leading to reduced power ratings of STATCOMs [27]. The resultant STATCOMs are essentially hybrid power quality conditioning equipment consisting of passive and active approaches.

3.5 Active Power Filters (APFs)

Active power filters (APFs) are notable power quality conditioning equipment that aims to address current quality problems. As the name suggests, APFs mainly compensate for current harmonics. In addition, they can achieve reactive current or power compensation as well as imbalance mitigation. When solely used for reactive power compensation, APFs become STATCOMs (or DSTATCOMs). In this sense, STATCOMs can be regarded as special APFs. As compared to passive filters, APFs are more compact and flexible, as they can compensate harmonics as programed. Recent years witness progress in the developments of APFs.

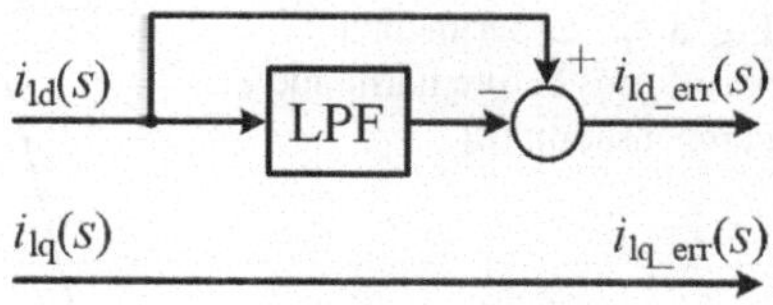

Fig. 3.41 Block diagram of *dq*0-frame current error measurement units

3.5.1 Fundamental Principles

Regarding hardware and topologies, APFs and STATCOMs are almost identical. Therefore, the schematic of Fig. 3.28 also applies to APFs. As harmonics cause more voltage drops than reactive currents, high-order interface filters (such as *LCL* filters) are preferable to avoid over modulation [28]. Simply speaking, APFs measure current errors (including current harmonics, reactive current, and imbalanced currents) from source or load currents and then inject current errors with opposite phase angles to power grids. As a result, grid currents become standard and sinusoidal. To compensate harmonics, APFs should incorporate multiple resonant or repetitive controllers, similar to those in DVRs [4].

3.5.2 Current Error Measurement Units

As introduced in Sect. 3.3, the voltage error measurement through frame rotations is effective and popular. Similarly, we often use frame rotations for current error measurements. For illustration, Fig. 3.41 shows the block diagram of *dq*0-frame current error measurement units, where the load current *dq*-frame components serve as inputs. The output error signals act as the references of current controllers.

3.5.3 Results

Figures 3.42, 3.43 and 3.44 demonstrate the experimental results of nonlinear loads before compensation, compensated by an *LCL*-filtered half-bridge APF, and compensated by an *LCCL*-filtered half-bridge APF, respectively [8]. In the experiments, grid voltages were emulated by a programmable ac voltage source (Chroma 61705).

As shown in Fig. 3.42, the gird current is distorted before compensation, which features a current THD of 33.1%. Such current distortions will distort grid voltages if line impedances are taken into consideration.

After compensated by the *LCL*-filtered APF, the grid current becomes more sinusoidal and in phase with the grid voltage. Correspondingly, the current THD reduces to 3.2%. However, there is an observable current spike around the switching frequency f_{sw} due to the limited attenuation of *LCL* filters.

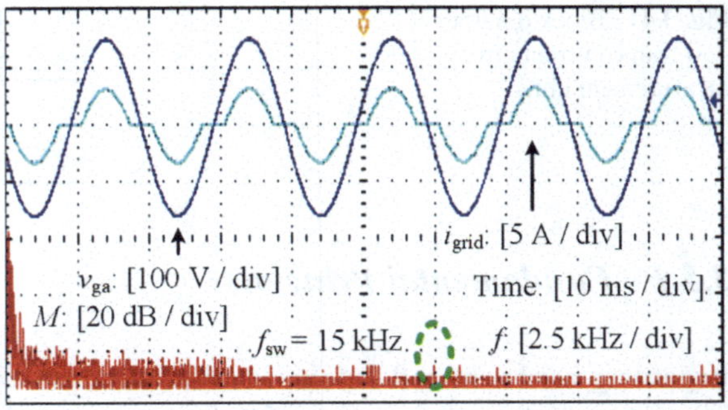

Fig. 3.42 Experimental waveforms before harmonic compensation [8]

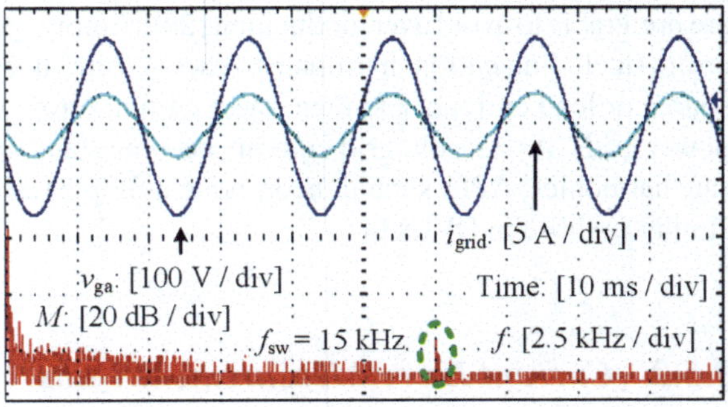

Fig. 3.43 Experimental waveforms of harmonic compensation by an *LCL*-filtered APF [8]

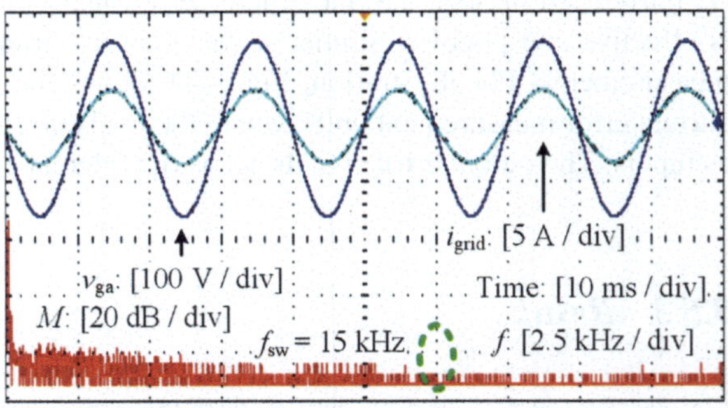

Fig. 3.44 Experimental waveforms of harmonic compensation by an *LCCL*-filtered APF [8]

In Fig. 3.44, we notice that the switching harmonics are sufficiently attenuated by the *LCCL*-filtered APF, leading to the compliance of grid codes. This result demonstrates the stronger harmonic attenuation ability of *LCCL* filters. It should be remarked that we refer to paralleled APFs in this subsection, while series APFs are denoted as DVRs in this book.

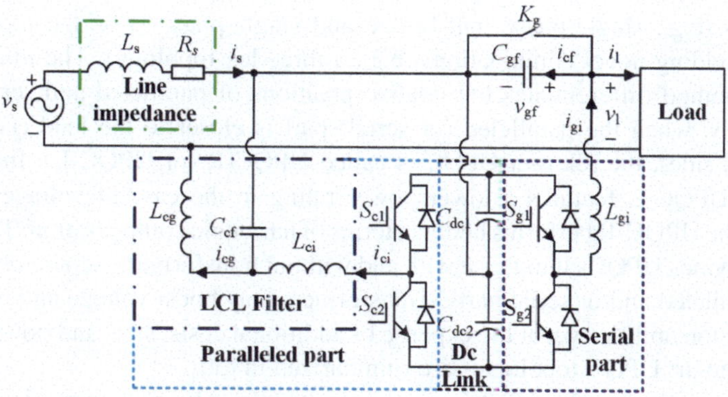

Fig. 3.45 Schematic of half-bridge transformerless UPQCs

3.6 Unified Power Quality Conditioners (UPQCs)

Unified power quality conditioners (UPQCs) combine the functions of DVRs and APFs, thus targeting at all power quality problems, including voltage amplitude deviations, voltage harmonics, voltage imbalances, reactive currents, current harmonics, and imbalanced currents, etc. As such, UPQCs are the most powerful power quality conditioning equipment [29].

3.6.1 *Fundamental Principles*

The UPQC consists of two parts—an APF and a DVR, which share a common dc link. As compared to a direct combination of one APF and one DVR, the UPQC features smaller size, lower weight and costs, and simpler control. In essence, UPQCs are ac-dc-ac power converters, where the two power stages (i.e., ac-dc and dc-ac power stages) corresponding to APFs and DVRs are generally different. As mentioned, APFs (or DVRs) are in parallel to (or in series with) loads. Therefore, the two parts of UPQCs are named as paralleled and serial parts, respectively.

In terms of topologies, we can combine all APF and DVR topologies into those of UPQCs as long as their dc links are identical. To classify topologies, the first means refers to the nature of ac-dc power converters, leading to two types—voltage and current source converters. Current source converters feature large dc inductors, which are often bulky and expensive. As for voltage sources, their dc links can be fed by batteries, external converters, or capacitors (or ultracapacitors). In comparison, dc capacitors require dedicated voltage controllers. The second method classifies UPQC topologies according to the phases of individual converters. As mentioned, single-phase half-bridge, single-phase H-bridge, three-phase three-line (with half-bridge or H-bridge), three-phase four-line topologies (with half-bridge or H-bridge) are common options. It should be remembered that the two parts can feature different topologies (e.g., single-phase half-bridge and single-phase H-bridge topologies), thereby yielding novel combinations (e.g., a three-leg topology). The third classification method differentiates the relative positions of paralleled and serial parts. Specifically, when the paralleled (or serial) part is closer to the load (i.e., in the right-hand side), the relevant UPQC is called UPQC-R (or UPQC-L). In comparison, the UPQC-L features a lower power rating in the case of voltage sags. In contrast, the UPQC-R performs better in terms of harmonic compensation. The fourth method divides UPQCs into those with and without transformers, separately applied to the paralleled and/or serial parts. Transformers may boost voltage and/or current compensation abilities yet at the expense of additional costs, size, and power losses. State-of-the-art UPQC topologies are summarized in [30].

As compared with other UPQCs, single-phase half-bridge transformerless UPQCs feature the simplest structure, which comprise only four active switches and two split dc capacitors. Figure 3.45 shows the schematic of half-bridge UPQCs, where the paralleled part ties to the grid through an *LCL* filter, while the serial part uses an *LC* filter as the grid interface. Generally, the paralleled part resolves current quality problems, meanwhile the serial part deals with voltage quality problems.

3.6.2 Modeling and Control

After dc voltages are well regulated, we can control UPQCs by separately controlling their paralleled parts as APFs and serial parts as DVRs. Notably, the dc voltage regulation and voltage balance schemes are detailed in [4, 31].

There is a chance of coordination through the two UPQC parts for power rating reduction. In the case of voltage sags, conventional UPQCs exploit only serial parts to compensate voltage sags. Instead, we may use serial parts to output a certain amount of reactive power and also paralleled parts for reactive power compensation, leading to a reduction of overall power rating [32, 33]. The price paid for size reduction is the phase angle difference between grid and load voltages. Generally, serial and paralleled parts are controlled as APFs and DVRs, respectively. However, the opposite control methodology is also possible (called direct control).

3.6.3 Results

This subsection provides the experimental results of a half-bridge UPQC with the standard controller based on the system parameters in Table 3.5.

Figure 3.46 shows the experimental results of UPQCs before operation, where the grid current i_s is distorted by the load current i_l with a THD of 31.2%.

Figure 3.47 presents the experimental results of UPQCs with load current compensation. After compensation, the grid current THD reduces from 31.2 to 2.8%, leading to the compliance of grid codes.

Table 3.5 UPQC system parameters

Descriptions	Symbols	Values
Dc capacitance	C_{dc}	4.7 mF
DC-link voltage reference	V_{dc_ref}	225 V
Grid voltage reference (rms)	V_{g_ref}	110 V
Fundamental frequency	f_o	50 Hz
Switching/sampling frequency	f_{sw}/f_s	15 kHz

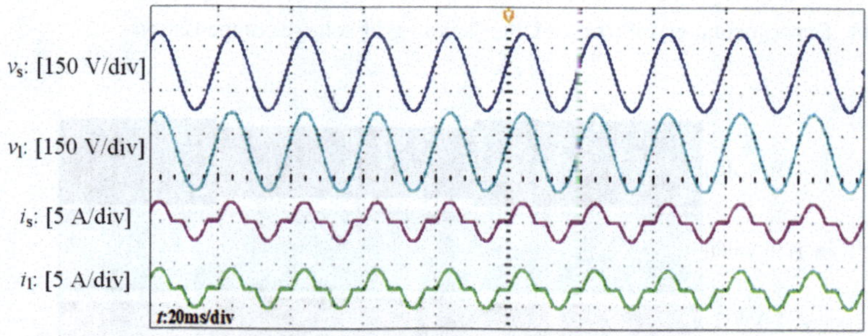

Fig. 3.46 Experimental waveforms of UPQCs without compensation

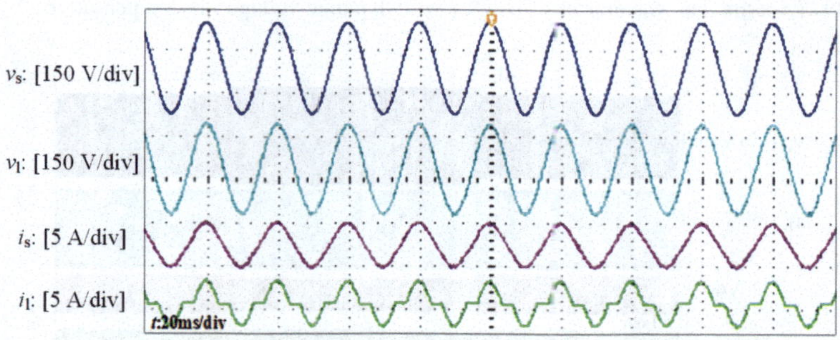

Fig. 3.47 Experimental waveforms of UPQCs with load current compensation

Figure 3.48 demonstrates the experimental results of UPQCs with grid voltage compensation. In this case, the grid voltage is contaminated with 10% 5th and 10% 7th harmonics, giving a THD of 14.5%. Meanwhile, the load current THD increases into 44.2% due to the distorted grid voltage before compensation. Thanks to the compensation of UPQCs, both the load voltage and grid current waveforms become sinusoidal. Correspondingly, their THDs reduce to 2.2% and 3.4%, respectively. Clearly, UPQCs exhibit satisfactory steady-state performances.

Figures 3.49 and 3.50 demonstrate the dynamic performances of UPQCs, where a grid voltage sag and a grid voltage swell of 4 s are registered, respectively. Noticeably, UPQCs manage to maintain standard load voltages regardless of grid voltages. These experimental results clearly verify the effectiveness of UPQCs.

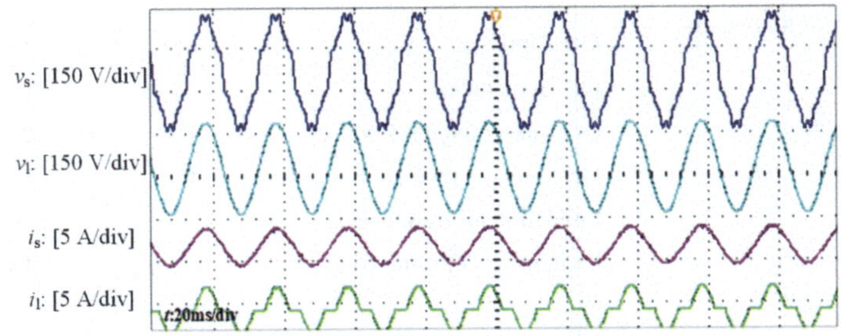

Fig. 3.48 Experimental waveforms of UPQCs with grid voltage compensation

Fig. 3.49 Experimental waveforms of UPQCs with dynamic voltage sag compensation

Fig. 3.50 Experimental waveforms of UPQCs with dynamic voltage swell compensation

In this chapter, we have investigated state-of-the-art solutions to power quality problems in more-electronics power systems. The chapter has discussed passive solutions for voltage amplitude, reactive power, and harmonic compensation. Importantly, we have detailed active solutions, namely, power electronics-based power conditioning equipment, including uninterruptable power supplies (UPSs), dynamic voltage restorers (DVRs), static compensators (STATCOMs), active power filters (APFs), and unified power quality conditioners (UPQCs). Their fundamental principles, modeling and control, and results have been highlighted.

References

1. Kundur P (1994) Power system stability and control. McGraw-Hill, New York, NY, USA
2. Dugan RC, McGranaghan MF, Santoso S, Beaty HW (2003) Electrical power systems quality. McGraw-Hill, 2 Penn Plaza, New York, NY, USA
3. Fang J, Lin P, Li H, Yang Y, Tang Y (2019) An improved virtual inertia control for three-phase voltage source converters connected to a weak grid. IEEE Trans Power Electron 34(9):8660–8670
4. Fang J, Xiao G, Yang X, Tang Y (2017) Parameter design of a novel series-parallel-resonant *LCL* filter for single-phase half-bridge active power filters. IEEE Trans Power Electron 32(1):200–217
5. Fang J, Li X, Yang X, Tang Y (2017) An integrated trap–*LCL* filter with reduced current harmonics for grid-connected converters under weak grid conditions. IEEE Trans Power Electron 32(11):8446–8457
6. Srianthumrong S, Akagi H (2003) Medium-voltage transformerless ac/dc power conversion system consisting of a diode rectifier and a shunt hybrid filter. IEEE Trans Ind Appl 39(3):874–882
7. Wu W, He Y, Blaabjerg F (2012) An *LLCL* power filter for single-phase grid-tied inverter. IEEE Trans Power Electron 27(2):782–789
8. Fang J, Xiao G, Zhang Y (2015) An LCCL filter and its application to a half-bridge APF. In: Presented at the Proceedings of the ICPE 2015–ECCE Asia, Seoul, Korea, pp 2566–2573
9. Zhang X, Zhu H, Li H, Yu C (2015) An *LCL-LC* filter for grid-connected converter: topology, parameter, and analysis. IEEE Trans Power Electron 30(9):5067–5077
10. Wu W et al (2014) A modified *LLCL* filter with the reduced conducted EMI noise. IEEE Trans Power Electron 29(7):3393–3402
11. Xu J, Yang J, Ye J, Zhang Z, Shen A (2014) An *LTCL* filter for three-phase grid-connected converters. IEEE Trans Power Electron 29(8):4322–4338
12. Li X, Fang J, Lin P, Tang Y (2018) Active magnetic decoupling for improving the performance of integrated *LCL*-filters in grid-connected converters. IEEE Trans Ind Electron 65(2):1367–1376
13. Wu W, He Y, Lin Z, Blaabjerg F, Chung HS-H (2015) A simple differential mode EMI suppressor for the *LLCL*-filter-based single-phase grid-tied transformerless inverter. IEEE Trans Ind Electron 62(7):4141–4147
14. Wu W, Sun Y, Lin Z, Tang T, Blaabjerg F, Chung HS-H (2014) A new *LCL*-filter with in-series parallel resonant circuit for single-phase grid-tied inverter. IEEE Trans Ind Electron 61(9):4640–4644
15. Holmes DG, Lipo TA (2003) Pulse width modulation for power converters: principles and practice. Wiley, Hoboken, NJ
16. Erickson RW, Maksimovic D (2001) Fundamentals of power electronics. Springer, New York, NY, USA
17. Liserre M, Blaabjerg F, Hansen S (2005) Design and control of an *LCL* filter-based three-phase active rectifier. IEEE Trans Ind Appl 41(5):1281–1291
18. Fang J, Li H, Tang Y (2017) A magnetic integrated *LLCL* filter for grid-connected voltage source converters. IEEE Trans Power Electron 32(3):1725–1730
19. Fang J, Li X, Tang Y, Li H (2018) Design of virtual synchronous generators with enhanced frequency regulation and reduced voltage distortions. In: Presented at the Proceedings of the IEEE APEC, San Antonio, TEXAS, USA, pp 1412–1419
20. Sype DM, Gusseme KD, Belie FM, Bossche AP, Melkebeek JA (2006) Small-signal *z*-domain analysis of digitally controlled converters. IEEE Trans Power Electron 21(2):470–478
21. Kai Z, Yong K, Jian X, Jian C (2003) Direct repetitive control of SPWM inverter for UPS purpose. IEEE Trans Power Electron 18(3):784–792
22. García-Cerrada A, Pinzón-Ardila O, Feliu-Batlle V, Roncero-Sánchez P, García-González P (2007) Application of a repetitive controller for a three-phase active power filter. IEEE Trans Power Electron 22(1):237–246

23. Li YW, Vilathgamuwa DM, Blaabjerg F, Loh PC (2007) A robust control scheme for medium-voltage-level DVR implementation. IEEE Trans Power Electron 54(4):2249–2261
24. Fang J, Zhang R, Li H, Tang Y (2019) Frequency derivative-based inertia enhancement by grid-connected power converters with a frequency-locked-loop. IEEE Trans Smart Grid 10(5):4918–4927
25. Peng FZ, Lai J (1996) Dynamic performance and control of a static var generator using cascade multilevel inverters. IEEE Trans Ind Appl 32(3):509–517
26. Wang J, Yan JD, Jiang L, Zou J (2016) Delay-dependent stability of single-loop controlled grid-connected inverters with *LCL* filters. IEEE Trans Power Electron 31(1):743–757
27. Fang J, Li X, Tang Y (2017) A novel *LCL*-filtered single-phase half-bridge distributed static compensator with DC-link filter capacitors and reduced passive component parameters. In: Proceedings of the IEEE APEC, Tampa, Florida, USA, pp 3279–3285
28. Tang Y, Loh PC, Wang P, Choo FH, Gao F, Blaabjerg F (2012) Generalized design of high performance shunt active power filter with output *LCL* filter. IEEE Trans Ind Electron 59(3):1443–1452
29. Fujita H, Akagi H (1998) The unified power quality conditioner: the integration of series- and shunt-active filters. IEEE Trans Power Electron 13(2):315–322
30. Khadkikar V (2012) Enhancing electric power quality using UPQC: a comprehensive overview. IEEE Trans Power Electron 27(5):2284–2297
31. Fang J, Li Z, Goetz SM (2020) Multilevel converters with symmetrical half-bridge submodules and sensorless voltage balance. IEEE Trans Power Electron (in press)
32. Khadkikar V, Chandra A (2008) A new control philosophy for a unified power quality conditioner (UPQC) to coordinate load-reactive power demand between shunt and series inverters. IEEE Trans Power Del 23(4):2522–2534
33. Khadkikar V, Chandra A (2011) UPQC-S: a novel concept of simultaneous voltage sag/swell and load reactive power compensations utilizing series inverter of UPQC. IEEE Trans Power Electron 26(9):2414–2425

Chapter 4
Stability Problems

4.1 Background of Stability Problems

Stable and secure operations are the priority of more-electronics power systems. As mentioned in Chap. 1, voltage and current waveforms of unstable systems will diverge after being disturbed, making system performances meaningless. In stable more-electronics power systems, all the components and devices should operate stably. Therefore, stability of more-electronics power systems is twofold. First, we should guarantee stable operations of individual power converters in the converter level. Second, the system-level stability must be ensured. Correspondingly, stability problems are classified into converter-level and system-level problems.

Before introducing stability problems, we should familiar ourselves with the elements in more-electronics power systems. In Chap. 1, we have already introduced the major elements and their models of conventional power systems. In more-electronics power systems, grid-tied power converters are the most salient units. In Chap. 3, we have covered the plant models of grid-tied power converters (as power quality conditioning equipment). Further, we will classify grid-tied power converters in terms of stability analysis and build their models which will greatly be helpful for exploration of stability problems.

4.1.1 Classification of Grid-Tied Converters

Grid-tied converters (GTCs) can be classified according to various standards. In this subsection, we briefly classify GTCs according to their topologies and control. The major objective of such a classification is to facilitate the introduction of stability problems. As such, the presented classification may not be unique.

As mentioned, standard GTC topologies include single-phase half-bridge, single-phase H-bridge, three-phase three-line (half-bridge or full-bridge), and three-phase

J. Fang, *More-Electronics Power Systems: Power Quality and Stability*, Power Systems, https://doi.org/10.1007/978-981-15-8590-6_4

four-line (half-bridge or H-bridge). In this subsection, we focus on the standard three-phase three-line half-bridge topology. On top of that, transformers are removed for simplicity. Besides, we consider the commonly used two-level converters instead of multilevel converters. In terms of control, we employ the synchronous *dq*0-frame control of GTCs due to the clear physical meanings of current and voltage signals. Under these conditions, we further classify GTCs.

Figure 4.1 shows a general schematic of GTCs, where the power grid is modeled as a series connection of line inductors L_s and grid voltage sources v_{sabc}. The GTC is tied to the grid through an *LCL* filter, which consists of converter-side inductors L_{gi}, filter capacitors C_{gf}, and grid-side inductors L_{gg}.

From the perspective of power grids, GTCs are simplified into either ac current or voltage sources. When controlled as ac current sources, GTCs follow existing grid voltages for current and power regulation, hence named as grid-following GTCs [1]. Notably, grid-following GTCs cannot operate alone under islanded conditions. Alternatively, when GTCs are controlled as ac voltage sources (defined as grid-forming GTCs), they can form grids under islanded conditions [1]. From the above discussion, we classify GTCs into two basic types—grid-following and grid-forming GTCs. In terms of control, grid-following GTCs regulate grid-injected currents i_{ggabc}, while grid-forming GTCs control ac capacitor voltages v_{gfabc}. We employ *LCL* filters for harmonic filtering and local voltage support.

Another classification criterion lies in the dc link. In the case of regulated dc links, e.g., fed by batteries or other converters, GTCs should regulate active power (including the state of charge of batteries). Notably, GTCs with regulated dc links exhibit a dc grid-following feature. In comparison, GTCs with unregulated dc capacitors must control their dc voltages through dedicated controllers, as mentioned in Sect. 3.5. When dc voltages are regulated, GTCs possess a dc grid-forming capability with indirectly regulated active power. It is noted from the above discussions that the nature of dc links divides GTCs into another two types. In conclusion, we classify GTCs into four types—ac grid-following + dc grid-following (Type I), ac grid-following + dc grid-forming (Type II), ac grid-forming + dc grid-following (Type III), and ac grid-forming + dc grid-forming (Type IV). We summarize the classification and fundamental objectives of GTCs in Table 4.1.

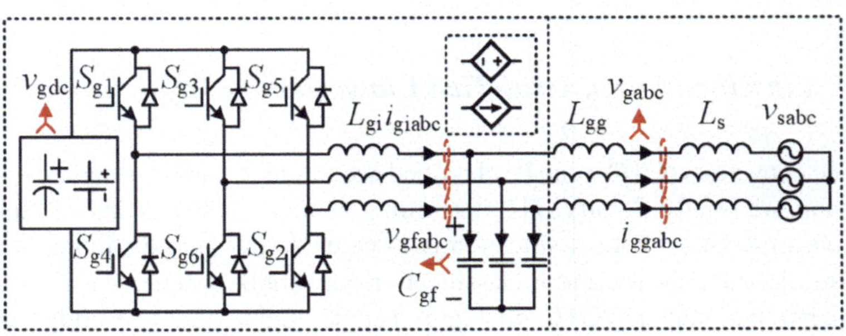

Fig. 4.1 General schematic of grid-tied converters

Table 4.1 Classification of grid-tied converters

GTCs	Explanations	Inner controllers	Outer controllers
Type I	Ac follow + dc follow	Ac currents	Active power
Type II	Ac follow + dc form	Ac currents	Dc voltage
Type III	Ac form + dc follow	Ac voltages	Active power
Type IV	Ac form + dc form	Ac voltages	Dc voltage

4.1.2 Modeling of Grid-Tied Converters

After introducing the GTC classification, we continue to build GTC models, which will lay a foundation for the subsequent study of stability problems. In Chap. 3, we mainly focus on block diagram models and classic controllers. Instead, we build state-space models of GTCs in this part. Let us start with the system plant model.

System Plant Model

We have already introduced the model of GTC system plants in Chap. 3 (see Fig. 3.30). With the resistors ignored, we redraw the block diagram of system plants in Fig. 4.2. As a step further, we derive the relationships among steady-state variables (represented by capital letters) as

$$\begin{bmatrix} V_{\mathrm{gid}} \\ V_{\mathrm{giq}} \end{bmatrix} = \begin{bmatrix} V_{\mathrm{gfd}} - \omega_0 L_{\mathrm{gi}} I_{\mathrm{giq}} \\ V_{\mathrm{gfq}} + \omega_0 L_{\mathrm{gi}} I_{\mathrm{gid}} \end{bmatrix}, \tag{4.1}$$

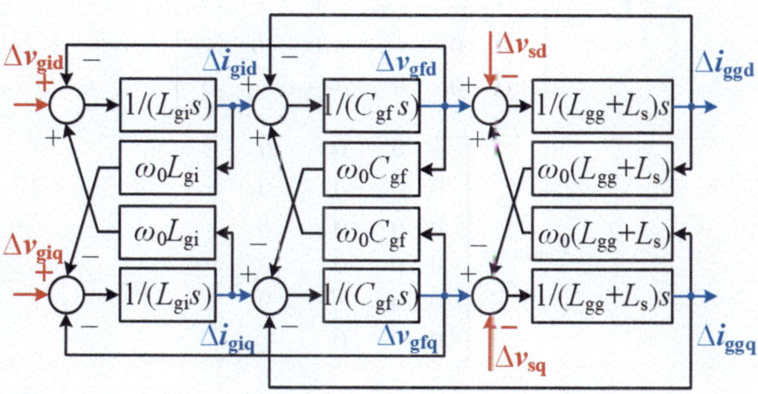

Fig. 4.2 Block diagram of system plants

$$\begin{bmatrix} V_{\mathrm{gfd}} \\ V_{\mathrm{gfq}} \end{bmatrix} = \begin{bmatrix} V_{\mathrm{sd}} - \omega_0 (L_{\mathrm{gg}} + L_{\mathrm{s}}) I_{\mathrm{ggq}} \\ V_{\mathrm{sq}} + \omega_0 (L_{\mathrm{gg}} + L_{\mathrm{s}}) I_{\mathrm{ggd}} \end{bmatrix}, \tag{4.2}$$

$$\begin{bmatrix} V_{\mathrm{gd}} \\ V_{\mathrm{gq}} \end{bmatrix} = \begin{bmatrix} \frac{V_{\mathrm{sd}} L_{\mathrm{gg}}}{L_{\mathrm{gg}}+L_{\mathrm{s}}} + \frac{V_{\mathrm{gfd}} L_{\mathrm{s}}}{L_{\mathrm{gg}}+L_{\mathrm{s}}} \\ \frac{V_{\mathrm{sq}} L_{\mathrm{gg}}}{L_{\mathrm{gg}}+L_{\mathrm{s}}} + \frac{V_{\mathrm{gfq}} L_{\mathrm{s}}}{L_{\mathrm{gg}}+L_{\mathrm{s}}} \end{bmatrix}, \tag{4.3}$$

$$\text{and} \begin{bmatrix} I_{\mathrm{gid}} \\ I_{\mathrm{giq}} \end{bmatrix} = \begin{bmatrix} I_{\mathrm{ggd}} - \omega_0 C_{\mathrm{gf}} V_{\mathrm{gfq}} \\ I_{\mathrm{ggq}} + \omega_0 C_{\mathrm{gf}} V_{\mathrm{gfd}} \end{bmatrix}, \tag{4.4}$$

where V_{gd} and V_{gq} refer to the point of common coupling (PCC) voltages v_{gabc} in Fig. 4.1. According to Fig. 4.2, we build the state-space model of system plants as

$$\begin{aligned} \tfrac{\mathrm{d}}{\mathrm{d}t} \Delta \mathbf{x}_{\mathbf{plant}} &= \mathbf{A}_{\mathbf{plant}} \Delta \mathbf{x}_{\mathbf{plant}} + \mathrm{B}_{\mathbf{plant}} \Delta \mathbf{u}_{\mathbf{plant}} \\ \Delta \mathbf{y}_{\mathbf{plant}} &= \mathbf{I} \Delta \mathbf{x}_{\mathbf{plant}} \end{aligned}, \tag{4.5}$$

where **I** denotes an identity matrix. The state vector $\Delta \mathbf{x}_{\mathbf{plant}}$, input vector $\Delta \mathbf{u}_{\mathbf{plant}}$, and output vector $\Delta \mathbf{y}_{\mathbf{plant}}$ are expressed as

$$\begin{aligned} \Delta \mathbf{x}_{\mathbf{plant}} &= \begin{bmatrix} \Delta i_{\mathrm{gid}} & \Delta i_{\mathrm{giq}} & \Delta v_{\mathrm{gfd}} & \Delta v_{\mathrm{gfq}} & \Delta i_{\mathrm{ggd}} & \Delta i_{\mathrm{ggq}} \end{bmatrix}^{\mathrm{T}} \\ \Delta \mathbf{u}_{\mathbf{plant}} &= \begin{bmatrix} \Delta v_{\mathrm{gid}} & \Delta v_{\mathrm{giq}} & \Delta v_{\mathrm{sd}} & \Delta v_{\mathrm{sq}} \end{bmatrix}^{\mathrm{T}} \\ \Delta \mathbf{y}_{\mathbf{plant}} &= \begin{bmatrix} \Delta i_{\mathrm{gid}} & \Delta i_{\mathrm{giq}} & \Delta v_{\mathrm{gfd}} & \Delta v_{\mathrm{gfq}} & \Delta i_{\mathrm{ggd}} & \Delta i_{\mathrm{ggq}} \end{bmatrix}^{\mathrm{T}} \end{aligned}. \tag{4.6}$$

In addition, $\mathbf{A}_{\mathbf{plant}}$ and $\mathbf{B}_{\mathbf{plant}}$ in (4.5) take the forms of

$$\mathbf{A}_{\mathbf{plant}} = \begin{bmatrix} 0 & \omega_0 & \frac{-1}{L_{\mathrm{gi}}} & 0 & 0 & 0 \\ -\omega_0 & 0 & 0 & \frac{-1}{L_{\mathrm{gi}}} & 0 & 0 \\ \frac{1}{C_{\mathrm{gf}}} & 0 & 0 & \omega_0 & \frac{-1}{C_{\mathrm{gf}}} & 0 \\ 0 & \frac{1}{C_{\mathrm{gf}}} & -\omega_0 & 0 & 0 & \frac{-1}{C_{\mathrm{gf}}} \\ 0 & 0 & \frac{1}{L_{\mathrm{gg}}+L_{\mathrm{s}}} & 0 & 0 & \omega_0 \\ 0 & 0 & 0 & \frac{1}{L_{\mathrm{gg}}+L_{\mathrm{s}}} & -\omega_0 & 0 \end{bmatrix} \text{ and} \tag{4.7}$$

$$\mathbf{B}_{\mathbf{plant}} = \begin{bmatrix} \frac{1}{L_{\mathrm{gi}}} & 0 & 0 & 0 \\ 0 & \frac{1}{L_{\mathrm{gi}}} & 0 & 0 \\ 0 & 0 & 0 & 0 \\ 0 & 0 & 0 & 0 \\ 0 & 0 & \frac{-1}{L_{\mathrm{gg}}+L_{\mathrm{s}}} & 0 \\ 0 & 0 & 0 & \frac{-1}{L_{\mathrm{gg}}+L_{\mathrm{s}}} \end{bmatrix}. \tag{4.8}$$

Phase-Locked-Loop (PLL) Model

Ac grid-following GTCs (Type I and Type II GTCs) require phase-locked loops (PLLs) to obtain the grid-voltage phase angle through the measurement of PCC voltages. There are various PLLs. The standard synchronous *dq*-frame PLL exploits a PI controller cascaded with an integrator to null the *q*-axis voltage [2]. PLLs output the phase angle reference of *abc*/*dq*0 transformations.

Figure 4.3 presents the model of synchronous-frame-based PLLs, where the detailed derivation can be referred to [3]. In Fig. 4.3, T_{pll} stands for the time constant of PLL filters. K_{pllp} and K_{plli} represent the proportional and integral gains of the PLL controller, respectively. $\Delta\omega_{\text{g}}$ and $\Delta\delta$ designate the perturbed frequency and phase angle given by PLLs.

We can implement the PLL model via various state-space equations. For example, a controllability canonical implementation is given as

$$\begin{aligned}\frac{\mathrm{d}}{\mathrm{d}t}\Delta\mathbf{x}_{\mathbf{pll}} &= \mathbf{A}_{\mathbf{pll}}\Delta\mathbf{x}_{\mathbf{pll}} + \mathrm{B}_{\mathbf{pll}}\Delta\mathbf{u}_{\mathbf{pll}} \\ \Delta\mathbf{y}_{\mathbf{pll}} &= \mathbf{C}_{\mathbf{pll}}\Delta\mathbf{x}_{\mathbf{pll}}\end{aligned}, \tag{4.9}$$

where

$$\begin{aligned}\Delta\mathbf{x}_{\mathbf{pll}} &= \left[\,\Delta x_{\text{pll1}}\ \Delta x_{\text{pll2}}\ \Delta x_{\text{pll3}}\,\right]^{\mathrm{T}} \\ \Delta\mathbf{u}_{\mathbf{pll}} &= \left[\,\Delta v_{\text{gfq}}\ \Delta v_{\text{sq}}\,\right]^{\mathrm{T}},\ \Delta\mathbf{y}_{\mathbf{pll}} = \left[\,\Delta\delta\ \Delta\omega_{\text{g}}\,\right]^{\mathrm{T}}\end{aligned}. \tag{4.10}$$

The relevant matrices are

$$\mathbf{A}_{\mathbf{pll}} = \begin{bmatrix} 0 & 1 & 0 \\ 0 & 0 & 1 \\ \frac{-V_{\text{gd}}K_{\text{plli}}}{T_{\text{pll}}} & \frac{-V_{\text{gd}}K_{\text{pllp}}}{T_{\text{pll}}} & \frac{-1}{T_{\text{pll}}} \end{bmatrix}, \tag{4.11}$$

$$\mathbf{B}_{\mathbf{pll}} = \begin{bmatrix} 0 & 0 \\ 0 & 0 \\ \frac{L_{\text{s}}}{L_{\text{gg}}+L_{\text{s}}} & \frac{L_{\text{gg}}}{L_{\text{gg}}+L_{\text{s}}} \end{bmatrix}, \text{ and} \tag{4.12}$$

$$\mathbf{C}_{\mathbf{pll}} = \begin{bmatrix} \frac{K_{\text{plli}}}{T_{\text{pll}}} & \frac{K_{\text{pllp}}}{T_{\text{pll}}} & 0 \\ 0 & \frac{K_{\text{plli}}}{T_{\text{pll}}} & \frac{K_{\text{pllp}}}{T_{\text{pll}}} \end{bmatrix}. \tag{4.13}$$

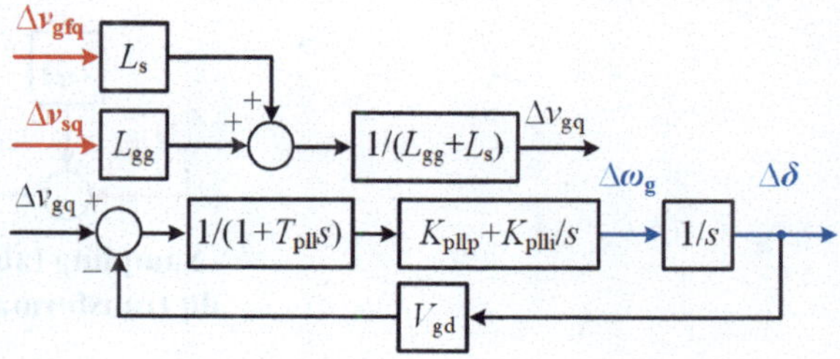

Fig. 4.3 Block diagram of standard synchronous-frame-based PLLs

Sampling and Update Models

The phase angle reference change $\Delta\delta$ (given by PLLs in Fig. 4.3) affects signal sampling and update through frame transformations [4]. After linearization, we present the block diagram representations of sampling and update in Fig. 4.4, where the subscript c describes control variables. Notably, control variables correspond to plant variables. However, control and plant variables are mapped to different frames due to frame rotations. We model the sampling part as

$$\Delta\mathbf{y}_{\mathbf{sample}} = \mathbf{I}\Delta\mathbf{x}_{\mathbf{sample}} + \mathbf{D}_{\mathbf{sample}}\Delta\mathbf{u}_{\mathbf{sample}}, \tag{4.14}$$

where

$$\begin{aligned} \Delta\mathbf{x}_{\mathbf{sample}} &= \left[\, \Delta i_{\mathrm{gid}}\ \Delta i_{\mathrm{giq}}\ \Delta v_{\mathrm{gfd}}\ \Delta v_{\mathrm{gfq}}\ \Delta i_{\mathrm{ggd}}\ \Delta i_{\mathrm{ggq}} \,\right]^{\mathrm{T}} \\ \Delta\mathbf{u}_{\mathbf{sample}} &= \Delta\delta \qquad\qquad\qquad\qquad\qquad\qquad\qquad \text{and} \\ \Delta\mathbf{y}_{\mathbf{sample}} &= \left[\, \Delta i_{\mathrm{gid\_c}}\ \Delta i_{\mathrm{giq\_c}}\ \Delta v_{\mathrm{gfd\_c}}\ \Delta v_{\mathrm{gfq\_c}}\ \Delta i_{\mathrm{ggd\_c}}\ \Delta i_{\mathrm{ggq\_c}} \,\right]^{\mathrm{T}} \end{aligned} \tag{4.15}$$

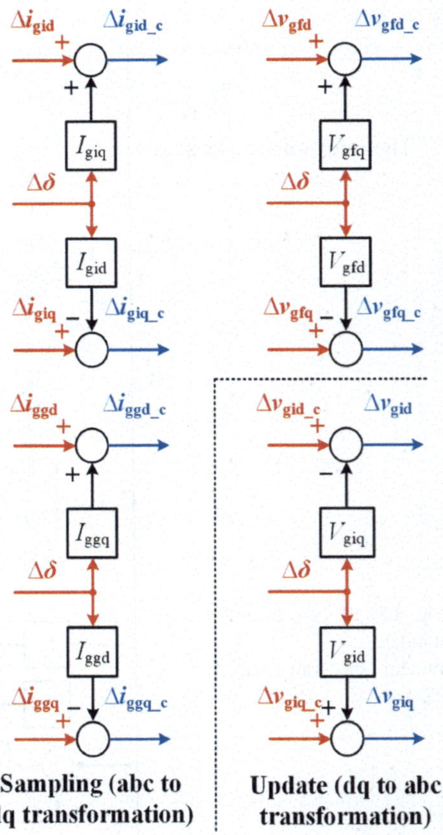

Fig. 4.4 Block diagrams of signal sampling and update

$$\mathbf{D}_{\mathbf{sample}} = \begin{bmatrix} I_{\mathrm{giq}} & -I_{\mathrm{gid}} & V_{\mathrm{gfq}} & -V_{\mathrm{gfd}} & I_{\mathrm{ggq}} & -I_{\mathrm{ggd}} \end{bmatrix}^{\mathrm{T}}. \tag{4.16}$$

The update part is modeled as

$$\Delta\mathbf{y}_{\mathbf{update}} = \mathbf{I}\Delta\mathbf{x}_{\mathbf{update}} + \mathbf{D}_{\mathbf{update}}\Delta\mathbf{u}_{\mathbf{update}}, \tag{4.17}$$

where

$$\begin{aligned} \Delta\mathbf{x}_{\mathbf{update}} &= \begin{bmatrix} \Delta v_{\mathrm{gid\_c}} & \Delta v_{\mathrm{giq\_c}} \end{bmatrix}^{\mathrm{T}} \\ \Delta\mathbf{u}_{\mathbf{update}} &= \Delta\delta,\ \Delta\mathbf{y}_{\mathbf{update}} = \begin{bmatrix} \Delta v_{\mathrm{gid}} & \Delta v_{\mathrm{giq}} \end{bmatrix}^{\mathrm{T}} \end{aligned} \text{ and} \tag{4.18}$$

$$\mathbf{D}_{\mathbf{update}} = \begin{bmatrix} -V_{\mathrm{giq}} & V_{\mathrm{gid}} \end{bmatrix}^{\mathrm{T}}. \tag{4.19}$$

Measurement Models

Active power, reactive power, and/or dc voltage measurements are necessary for different types of GTCs. Figure 4.5 shows the block diagram representations of measurements, where T_{p} and T_{q} denote the time constants of the low-pass filters of active and reactive power measurements, respectively. C_{gdc} refers to the dc-link capacitance. Note that active and reactive power are measured indirectly from the calculation of control variables (with the subscript c), while the dc voltage (without the subscript c) is directly measured [5].

For dc grid-following GTCs (Type I and Type III), we measure active and reactive power. In contrast, dc voltage and reactive power are measured for dc grid-forming GTCs (Type II and Type IV). Specifically, the active and reactive power measurement model is

$$\begin{aligned} \tfrac{\mathrm{d}}{\mathrm{d}t}\Delta\mathbf{x}_{\mathbf{pq\_measure}} &= \mathbf{A}_{\mathbf{pq\_measure}}\Delta\mathbf{x}_{\mathbf{pq\_measure}} + \mathbf{B}_{\mathbf{pq\_measure}}\Delta\mathbf{u}_{\mathbf{pq\_measure}} \\ \Delta\mathbf{y}_{\mathbf{pq\_measure}} &= \mathbf{I}\Delta\mathbf{x}_{\mathbf{pq\_measure}} \end{aligned}, \tag{4.20}$$

where

$$\begin{aligned} \Delta\mathbf{x}_{\mathbf{pq\_measure}} &= \begin{bmatrix} \Delta p_{\mathrm{g\_c}} & \Delta q_{\mathrm{g\_c}} \end{bmatrix}^{\mathrm{T}},\ \Delta\mathbf{y}_{\mathbf{pq\_measure}} = \begin{bmatrix} \Delta p_{\mathrm{g\_c}} & \Delta q_{\mathrm{g\_c}} \end{bmatrix}^{\mathrm{T}} \\ \Delta\mathbf{u}_{\mathbf{pq\_measure}} &= \begin{bmatrix} \Delta v_{\mathrm{gfd\_c}} & \Delta v_{\mathrm{gfq\_c}} & \Delta i_{\mathrm{ggd\_c}} & \Delta i_{\mathrm{ggq\_c}} \end{bmatrix}^{\mathrm{T}} \end{aligned}. \tag{4.21}$$

The matrices in (4.20) are derived as

$$\mathbf{A}_{\mathbf{pq\_measure}} = \begin{bmatrix} \frac{-1}{T_{\mathrm{p}}} & 0 \\ 0 & \frac{-1}{T_{\mathrm{q}}} \end{bmatrix} \text{ and} \tag{4.22}$$

$$\mathbf{B}_{\mathbf{pq\_measure}} = \begin{bmatrix} \frac{3I_{\mathrm{ggd}}}{2T_{\mathrm{p}}} & \frac{3I_{\mathrm{ggq}}}{2T_{\mathrm{p}}} & \frac{3V_{\mathrm{gfd}}}{2T_{\mathrm{p}}} & \frac{3V_{\mathrm{gfq}}}{2T_{\mathrm{p}}} \\ \frac{-3I_{\mathrm{ggq}}}{2T_{\mathrm{q}}} & \frac{3I_{\mathrm{ggd}}}{2T_{\mathrm{q}}} & \frac{3V_{\mathrm{gfq}}}{2T_{\mathrm{q}}} & \frac{-3V_{\mathrm{gfd}}}{2T_{\mathrm{q}}} \end{bmatrix}. \tag{4.23}$$

Fig. 4.5 Block diagrams of active power, reactive power, and dc-voltage measurements

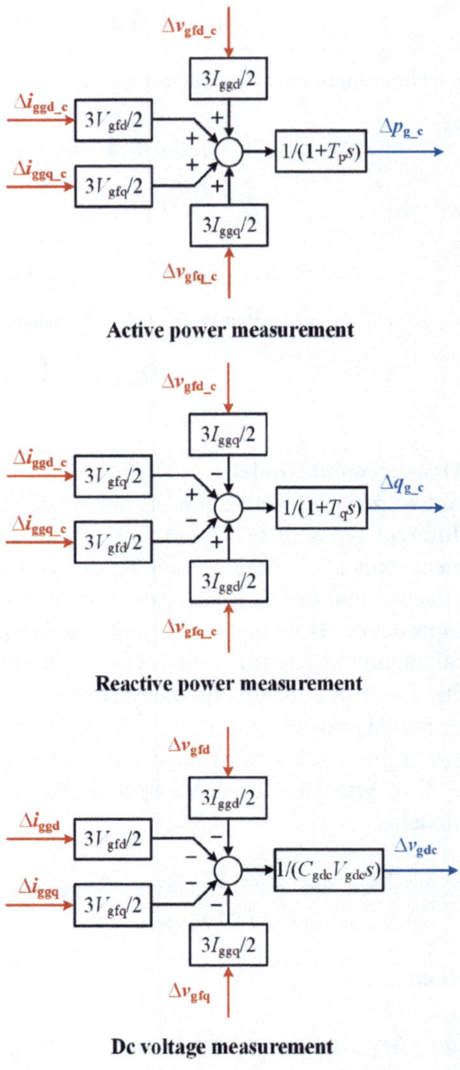

Referring to the block diagram of the dc voltage measurement in Fig. 4.5, we can replace its input plant variables (without the subscript c) by the corresponding control variables (with the subscript c). This is because the two cases lead to identical outputs Δv_{gdc}, as proven by the substitution of control variables into plant variables according to the relationships in Fig. 4.4. As a result, the input signals of dc voltage and reactive power measurements become identical. The resultant dc voltage and reactive power measurement part is modeled as

$$\begin{aligned}\frac{d}{dt}\Delta\mathbf{x}_{\mathbf{vdcq\_measure}} &= \mathbf{A}_{\mathbf{vdcq\_measure}}\Delta\mathbf{u}_{\mathbf{vdcq\_measure}} \\ &+ \mathbf{B}_{\mathbf{vdcq\_measure}}\Delta\mathbf{u}_{\mathbf{vdcq\_measure}} \\ \Delta\mathbf{y}_{\mathbf{vdcq\_measure}} &= \mathbf{I}\Delta\mathbf{x}_{\mathbf{vdcq\_measure}}\end{aligned}, \tag{4.24}$$

where

$$\begin{aligned}\Delta\mathbf{x}_{\mathbf{vdcq\_measure}} &= \left[\Delta v_{\mathrm{gdc}}\ \Delta q_{\mathrm{g\_c}}\right]^{\mathrm{T}}, \Delta\mathbf{y}_{\mathbf{vdcq\_measure}} = \left[\Delta v_{\mathrm{gdc}}\ \Delta q_{\mathrm{g\_c}}\right]^{\mathrm{T}} \\ \Delta\mathbf{u}_{\mathbf{vdcq\_measure}} &= \left[\Delta v_{\mathrm{gfd\_c}}\ \Delta v_{\mathrm{gfq\_c}}\ \Delta i_{\mathrm{ggd\_c}}\ \Delta i_{\mathrm{ggq\_c}}\right]^{\mathrm{T}}\end{aligned}, \tag{4.25}$$

$$\mathbf{A}_{\mathbf{vdcq\_measure}} = \begin{bmatrix} 0 & 0 \\ 0 & \frac{-1}{T_{\mathrm{q}}} \end{bmatrix} \text{ and} \tag{4.26}$$

$$\mathbf{B}_{\mathbf{vdcq\_measure}} = \begin{bmatrix} \frac{-3I_{\mathrm{ggd}}}{2C_{\mathrm{gdc}}V_{\mathrm{gdc}}} & \frac{-3I_{\mathrm{ggq}}}{2C_{\mathrm{gdc}}V_{\mathrm{gdc}}} & \frac{-3V_{\mathrm{gfd}}}{2C_{\mathrm{gdc}}V_{\mathrm{gdc}}} & \frac{-3V_{\mathrm{gfq}}}{2C_{\mathrm{gdc}}V_{\mathrm{gdc}}} \\ \frac{-3I_{\mathrm{ggq}}}{2T_{\mathrm{q}}} & \frac{3I_{\mathrm{ggd}}}{2T_{\mathrm{q}}} & \frac{3V_{\mathrm{gfq}}}{2T_{\mathrm{q}}} & \frac{-3V_{\mathrm{gfd}}}{2T_{\mathrm{q}}} \end{bmatrix}. \tag{4.27}$$

Ac Current Controller Models

In this part, we focus on the single-loop ac current controllers that regulate either grid or converter currents. Figure 4.6 illustrates the models of current controllers, where K_{cp} and K_{ci} represent the proportional and integral gains of current controllers, respectively. K_{id} denotes the current feedback coefficient. Additionally, the model of time delays is built in the s domain as $G_{\mathrm{d}}(s)$, which is a first-order Padé approximation of the ZOH and one sampling period delay, as given in Sect. 3.2.

We model grid-current controllers as

$$\begin{aligned}\frac{d}{dt}\Delta\mathbf{x}_{\mathbf{grid\_current}} &= \mathbf{A}_{\mathbf{grid\_current}}\Delta\mathbf{x}_{\mathbf{grid\_current}} + \mathbf{B}_{\mathbf{grid\_current}}\Delta\mathbf{u}_{\mathbf{grid\_current}} \\ \Delta\mathbf{y}_{\mathbf{grid\_current}} &= \mathbf{C}_{\mathbf{grid\_current}}\Delta\mathbf{x}_{\mathbf{grid\_current}} + \mathbf{D}_{\mathbf{grid\_current}}\Delta\mathbf{u}_{\mathbf{grid\_current}}\end{aligned}, \tag{4.28}$$

where

$$\begin{aligned}\Delta\mathbf{x}_{\mathbf{grid\_current}} &= \left[\Delta x_{\mathrm{cd}}\ \Delta x_{\mathrm{cq}}\ \Delta x_{\mathrm{dd}}\ \Delta x_{\mathrm{dq}}\right]^{\mathrm{T}} \\ \Delta\mathbf{u}_{\mathbf{grid\_current}} &= \left[\Delta i_{\mathrm{ggd\_ref}}\ \Delta i_{\mathrm{ggq\_ref}}\ \Delta i_{\mathrm{ggd\_c}}\ \Delta i_{\mathrm{ggq\_c}}\right]^{\mathrm{T}} \\ \Delta\mathbf{y}_{\mathbf{grid\_current}} &= \left[\Delta v_{\mathrm{gid\_c}}\ \Delta v_{\mathrm{giq\_c}}\right]^{\mathrm{T}}\end{aligned}. \tag{4.29}$$

We derive the relevant matrices as

$$\mathbf{A}_{\mathbf{grid\_current}} = \begin{bmatrix} 0 & 0 & 0 & 0 \\ 0 & 0 & 0 & 0 \\ \frac{8K_{\mathrm{ci}}}{3T_{\mathrm{s}}} & 0 & \frac{-4}{3T_{\mathrm{s}}} & 0 \\ 0 & \frac{8K_{\mathrm{ci}}}{3T_{\mathrm{s}}} & 0 & \frac{-4}{3T_{\mathrm{s}}} \end{bmatrix}, \tag{4.30}$$

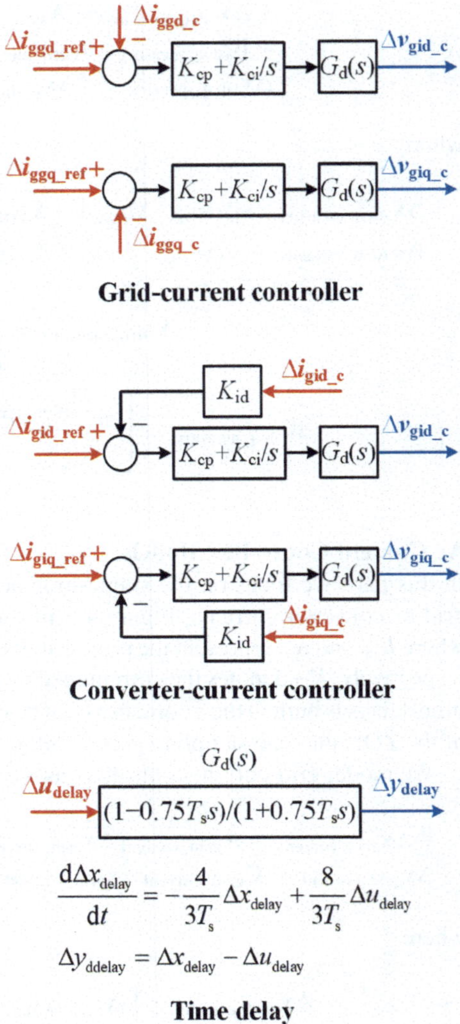

Fig. 4.6 Block diagrams of ac current controllers

$$\mathbf{B}_{\mathbf{grid\_current}} = \begin{bmatrix} 1 & 0 & -1 & 0 \\ 0 & 1 & 0 & -1 \\ \frac{8K_{\text{cp}}}{3T_s} & 0 & \frac{-8K_{\text{cp}}}{3T_s} & 0 \\ 0 & \frac{8K_{\text{cp}}}{3T_s} & 0 & \frac{-8K_{\text{cp}}}{3T_s} \end{bmatrix}, \tag{4.31}$$

$$\mathbf{C}_{\mathbf{grid\_current}} = \begin{bmatrix} -K_{\text{ci}} & 0 & 1 & 0 \\ 0 & -K_{\text{ci}} & 0 & 1 \end{bmatrix}, \text{ and} \tag{4.32}$$

$$\mathbf{D}_{\mathbf{grid\_current}} = \begin{bmatrix} -K_{\text{cp}} & 0 & K_{\text{cp}} & 0 \\ 0 & -K_{\text{cp}} & 0 & K_{\text{cp}} \end{bmatrix}. \tag{4.33}$$

The state variables Δx_{cd} (and Δx_{cq}) or Δx_{dd} (and Δx_{dq}) in (4.29) are caused by PI regulators or time delays, respectively. We model converter-current controllers as

$$\begin{aligned}\tfrac{\mathrm{d}}{\mathrm{d}t}\Delta\mathbf{x}_{\mathbf{con\_current}} &= \mathbf{A}_{\mathbf{con\_current}}\Delta\mathbf{x}_{\mathbf{con\_current}} + \mathrm{B}_{\mathbf{con\_current}}\Delta\mathbf{u}_{\mathbf{con\_current}} \\ \Delta\mathbf{y}_{\mathbf{con\_current}} &= \mathbf{C}_{\mathbf{con\_current}}\Delta\mathbf{x}_{\mathbf{con\_current}} + \mathbf{D}_{\mathbf{con\_current}}\Delta\mathbf{u}_{\mathbf{con\_current}}\end{aligned}, \tag{4.34}$$

where

$$\begin{aligned}\Delta\mathbf{x}_{\mathbf{con\_current}} &= \begin{bmatrix}\Delta x_{\mathrm{cd}} & \Delta x_{\mathrm{cq}} & \Delta x_{\mathrm{dd}} & \Delta x_{\mathrm{dq}}\end{bmatrix}^{\mathrm{T}} \\ \Delta\mathbf{u}_{\mathbf{con\_current}} &= \begin{bmatrix}\Delta i_{\mathrm{gid\_ref}} & \Delta i_{\mathrm{giq\_ref}} & \Delta i_{\mathrm{gid\_c}} & \Delta i_{\mathrm{giq\_c}}\end{bmatrix}^{\mathrm{T}} \\ \Delta\mathbf{y}_{\mathbf{con\_current}} &= \begin{bmatrix}\Delta v_{\mathrm{gid\_c}} & \Delta v_{\mathrm{giq\_c}}\end{bmatrix}^{\mathrm{T}}\end{aligned}. \tag{4.35}$$

The relevant matrices are

$$\mathbf{A}_{\mathbf{con\_current}} = \begin{bmatrix}0 & 0 & 0 & 0 \\ 0 & 0 & 0 & 0 \\ \frac{8K_{\mathrm{ci}}}{3T_{\mathrm{s}}} & 0 & \frac{-4}{3T_{\mathrm{s}}} & 0 \\ 0 & \frac{8K_{\mathrm{ci}}}{3T_{\mathrm{s}}} & 0 & \frac{-4}{3T_{\mathrm{s}}}\end{bmatrix}, \tag{4.36}$$

$$\mathbf{B}_{\mathbf{con\_current}} = \begin{bmatrix}1 & 0 & -K_{\mathrm{id}} & 0 \\ 0 & 1 & 0 & -K_{\mathrm{id}} \\ \frac{8K_{\mathrm{cp}}}{3T_{\mathrm{s}}} & 0 & \frac{-8K_{\mathrm{cp}}K_{\mathrm{id}}}{3T_{\mathrm{s}}} & 0 \\ 0 & \frac{8K_{\mathrm{cp}}}{3T_{\mathrm{s}}} & 0 & \frac{-8K_{\mathrm{cp}}K_{\mathrm{id}}}{3T_{\mathrm{s}}}\end{bmatrix}, \tag{4.37}$$

$$\mathbf{C}_{\mathbf{con\_current}} = \begin{bmatrix}-K_{\mathrm{ci}} & 0 & 1 & 0 \\ 0 & -K_{\mathrm{ci}} & 0 & 1\end{bmatrix}, \text{ and} \tag{4.38}$$

$$\mathbf{D}_{\mathbf{con\_current}} = \begin{bmatrix}-K_{\mathrm{cp}} & 0 & K_{\mathrm{cp}}K_{\mathrm{id}} & 0 \\ 0 & -K_{\mathrm{cp}} & 0 & K_{\mathrm{cp}}K_{\mathrm{id}}\end{bmatrix}. \tag{4.39}$$

Comparing (4.30) [and (4.32)] with (4.36) [and (4.38)], we find that **A** (and **C**) matrices of the two types of controllers are identical.

Ac Voltage Controller Model

Ac grid-forming GTCs (Types III and IV) necessitate ac voltage controllers, whose block diagram is shown in Fig. 4.7, where K_{vp} and K_{vi} denote the proportional and integral gains of PI controllers.

We model ac voltage controllers as

$$\begin{aligned}\tfrac{\mathrm{d}}{\mathrm{d}t}\Delta\mathbf{x}_{\mathbf{ac\_voltage}} &= \mathrm{B}_{\mathbf{ac\_voltage}}\Delta\mathbf{u}_{\mathbf{ac\_voltage}} \\ \Delta\mathbf{y}_{\mathbf{ac\_voltage}} &= \mathbf{C}_{\mathbf{ac\_voltage}}\Delta\mathbf{x}_{\mathbf{ac\_voltage}} + \mathbf{D}_{\mathbf{ac\_voltage}}\Delta\mathbf{u}_{\mathbf{ac\_voltage}}\end{aligned}, \tag{4.40}$$

in which

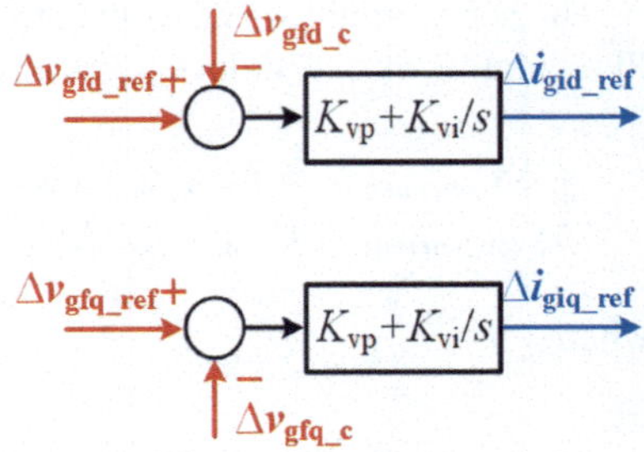

Fig. 4.7 Block diagram of ac voltage controllers

$$\begin{aligned} \Delta \mathbf{x}_{\mathbf{ac\_voltage}} &= \left[\Delta x_{vd} \ \Delta x_{vq} \right]^T, \Delta \mathbf{y}_{\mathbf{ac\_voltage}} = \left[\Delta i_{gid\_ref} \ \Delta i_{giq\_ref} \right]^T \\ \Delta \mathbf{u}_{\mathbf{ac\_voltage}} &= \left[\Delta v_{gfd\_ref} \ \Delta v_{gfq\_ref} \ \Delta v_{gfd\_c} \ \Delta v_{gfq\_c} \right]^T \end{aligned}. \tag{4.41}$$

We express the relevant matrices as

$$\mathbf{B}_{\mathbf{ac\_voltage}} = \begin{bmatrix} 1 & 0 & -1 & 0 \\ 0 & 1 & 0 & -1 \end{bmatrix}, \tag{4.42}$$

$$\mathbf{C}_{\mathbf{ac\_voltage}} = \begin{bmatrix} K_{vi} & 0 \\ 0 & K_{vi} \end{bmatrix}, \tag{4.43}$$

$$\mathbf{D}_{\mathbf{ac\_voltage}} = \begin{bmatrix} K_{vp} & 0 & -K_{vp} & 0 \\ 0 & K_{vp} & 0 & -K_{vp} \end{bmatrix}. \tag{4.44}$$

Dc Voltage Controller Model

Dc voltage controllers are necessary in dc grid-forming GTCs (i.e., Type II and Type IV GTCs). Specifically, for Type II GTCs, the dc voltage controller outputs a *d*-axis current reference Δi_{ggd_ref} for the ac current controller. In contrast, Type IV GTCs regulate the dc voltage through the change of the ac voltage phase angle $\Delta\delta$. Figure 4.8 illustrates the block diagrams of dc voltage controllers, where K_{dcp} and K_{dci} represent the proportional and integral gains of dc voltage controllers, respectively. Further, we model dc voltage controllers as.

$$\begin{aligned} \frac{d}{dt} \Delta \mathbf{x}_{\mathbf{dc\_voltage}} &= \mathrm{B}_{\mathbf{dc\_voltage}} \Delta \mathbf{u}_{\mathbf{dc\_voltage}} \\ \Delta \mathbf{y}_{\mathbf{dc\_voltage}} &= \mathbf{C}_{\mathbf{dc\_voltage}} \Delta \mathbf{x}_{\mathbf{dc\_voltage}} + \mathbf{D}_{\mathbf{dc\_voltage}} \Delta \mathbf{u}_{\mathbf{dc\_voltage}} \end{aligned}, \tag{4.45}$$

where

$$\begin{aligned} \Delta \mathbf{x}_{\mathbf{dc\_voltage}} &= \Delta x_{dc}, \Delta \mathbf{u}_{\mathbf{dc\_voltage}} = \left[\Delta v_{gdc\_ref} \ \Delta v_{gdc} \right]^T \\ \Delta \mathbf{y}_{\mathbf{dc\_voltage}} &= \Delta i_{ggd\_ref} \text{ or } \Delta\delta \end{aligned}. \tag{4.46}$$

The relevant matrices are

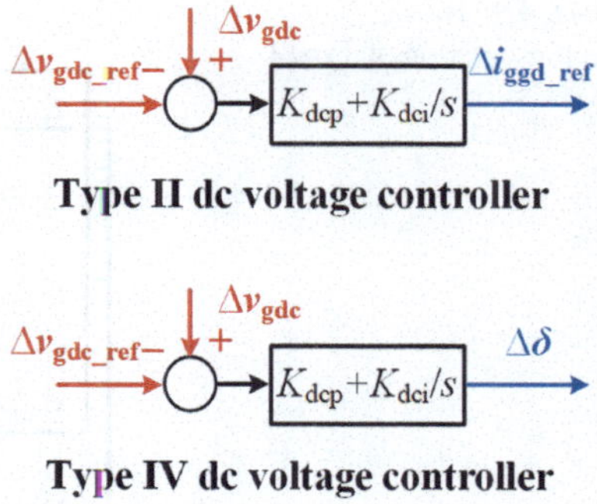

Fig. 4.8 Block diagrams of dc voltage controllers

$$\mathbf{B}_{\mathbf{dc\_voltage}} = \begin{bmatrix} -1 & 1 \end{bmatrix}, \tag{4.47}$$

$$\mathbf{C}_{\mathbf{dc\_voltage}} = K_{\mathrm{dci}}, \tag{4.48}$$

$$\mathbf{D}_{\mathbf{dc\_voltage}} = \begin{bmatrix} -K_{\mathrm{dcp}} & K_{\mathrm{dcp}} \end{bmatrix}. \tag{4.49}$$

It is noted from (4.46) that the output of dc voltage controllers can either be $\Delta i_{\mathrm{ggd\_ref}}$ or $\Delta\delta$ according to GTC types. Moreover, Δx_{dc} is introduced by the dc voltage PI regulator.

Overall System Models

So far, we have built the models of individual GTC parts. Next, we derive the overall system models. First, Fig. 4.9 shows the block representation of Type I GTCs, where the aforementioned blocks are interconnected with highlighted input and output signals.

We model Type I GTCs as

$$\begin{array}{l} \frac{\mathrm{d}}{\mathrm{d}t}\Delta\mathbf{x}_{\mathbf{TypeI}} = \mathbf{A}_{\mathbf{TypeI}}\Delta\mathbf{x}_{\mathbf{TypeI}} + \mathrm{B}_{\mathbf{TypeI}}\Delta\mathbf{u}_{\mathbf{TypeI}} \\ \Delta\mathbf{y}_{\mathbf{TypeI}} = \mathbf{C}_{\mathbf{TypeI}}\Delta\mathbf{x}_{\mathbf{TypeI}} \end{array}, \tag{4.50}$$

where

$$\begin{array}{l} \Delta\mathbf{x}_{\mathbf{TypeI}(1\times15)} = \begin{bmatrix} \Delta\mathbf{x}_{\mathbf{plant}} & \Delta\mathbf{x}_{\mathbf{grid\_current}} & \Delta\mathbf{x}_{\mathbf{pll}} & \Delta\mathbf{x}_{\mathbf{pq\_measure}} \end{bmatrix}^{\mathrm{T}} \\ \Delta\mathbf{u}_{\mathbf{TypeI}} = \begin{bmatrix} \Delta i_{\mathrm{ggd\_ref}} & \Delta i_{\mathrm{ggq\_ref}} & \Delta v_{\mathrm{sd}} & \Delta v_{\mathrm{sq}} \end{bmatrix}^{\mathrm{T}} \\ \Delta\mathbf{y}_{\mathbf{TypeI}} = \begin{bmatrix} \Delta i_{\mathrm{ggd}} & \Delta i_{\mathrm{ggq}} & \Delta\omega_{\mathrm{g}} & \Delta p_{\mathrm{g\_c}} & \Delta q_{\mathrm{g\_c}} \end{bmatrix}^{\mathrm{T}} \end{array}. \tag{4.51}$$

The output grid currents Δi_{ggd} and Δi_{ggq} should track their references $\Delta i_{\mathrm{ggd\_ref}}$ and $\Delta i_{\mathrm{ggq\_ref}}$, respectively. The grid voltages Δv_{sd} and Δv_{sq} are treated as disturbances. In addition, we reserve the frequency ($\Delta\omega_{\mathrm{g}}$) and power signals ($\Delta p_{\mathrm{g\_c}}$ and $\Delta q_{\mathrm{g\_c}}$) for external controllers.

The relevant matrices are

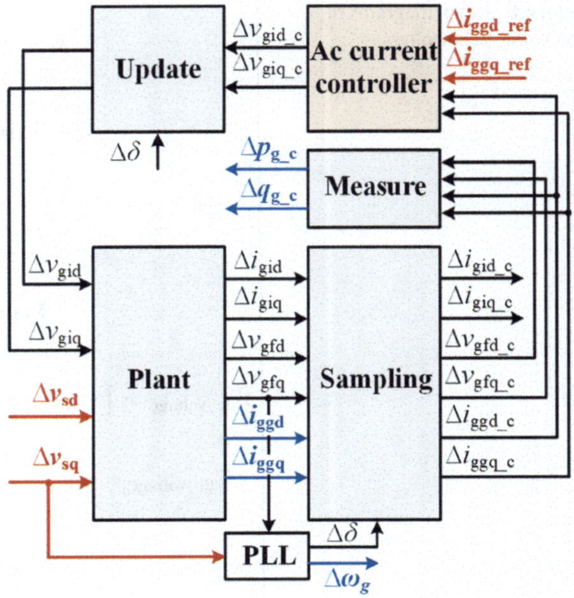

Fig. 4.9 Block representation of Type I GTCs

$$
\mathbf{A}_{\text{TypeI}} = \begin{bmatrix}
0 & \omega_0 & \frac{-1}{L_{gi}} & 0 & \frac{K_{cp}}{L_{gi}} & 0 & \frac{-K_{ci}}{L_{gi}} & 0 & \frac{1}{L_{gi}} & 0 & \frac{-V_{giq}K_{plli}+I_{ggq}K_{cp}K_{plli}}{T_{pll}L_{gi}} & \frac{-V_{giq}K_{pllp}+I_{ggq}K_{cp}K_{pllp}}{T_{pll}L_{gi}} & 0 & 0 & 0 \\
-\omega_0 & 0 & 0 & \frac{-1}{L_{gi}} & 0 & \frac{K_{cp}}{L_{gi}} & 0 & \frac{-K_{ci}}{L_{gi}} & 0 & \frac{1}{L_{gi}} & \frac{V_{gid}K_{plli}-I_{ggd}K_{cp}K_{plli}}{T_{pll}L_{gi}} & \frac{V_{gid}K_{pllp}-I_{ggd}K_{cp}K_{pllp}}{T_{pll}L_{gi}} & 0 & 0 & 0 \\
\frac{1}{C_{gf}} & 0 & 0 & \omega_0 & \frac{-1}{C_{gf}} & 0 & 0 & 0 & 0 & 0 & 0 & 0 & 0 & 0 & 0 \\
0 & \frac{1}{C_{gf}} & -\omega_0 & 0 & 0 & \frac{-1}{C_{gf}} & 0 & 0 & 0 & 0 & 0 & 0 & 0 & 0 & 0 \\
0 & 0 & \frac{1}{L_{gg}+L_s} & 0 & 0 & \omega_0 & 0 & 0 & 0 & 0 & 0 & 0 & 0 & 0 & 0 \\
0 & 0 & 0 & \frac{1}{L_{gg}+L_s} & -\omega_0 & 0 & 0 & 0 & 0 & 0 & 0 & 0 & 0 & 0 & 0 \\
0 & 0 & 0 & 0 & -1 & 0 & 0 & 0 & 0 & 0 & \frac{-I_{ggq}K_{plli}}{T_{pll}} & \frac{-I_{ggq}K_{pllp}}{T_{pll}} & 0 & 0 & 0 \\
0 & 0 & 0 & 0 & 0 & -1 & 0 & 0 & 0 & 0 & \frac{I_{ggd}K_{plli}}{T_{pll}} & \frac{I_{ggd}K_{pllp}}{T_{pll}} & 0 & 0 & 0 \\
0 & 0 & 0 & 0 & \frac{-8K_{cp}}{3T_s} & 0 & \frac{8K_{ci}}{3T_s} & 0 & \frac{-4}{3T_s} & 0 & \frac{-8I_{ggq}K_{cp}K_{plli}}{3T_sT_{pll}} & \frac{-8I_{ggq}K_{cp}K_{pllp}}{3T_sT_{pll}} & 0 & 0 & 0 \\
0 & 0 & 0 & 0 & 0 & \frac{-8K_{cp}}{3T_s} & 0 & \frac{8K_{ci}}{3T_s} & 0 & \frac{-4}{3T_s} & \frac{8I_{ggd}K_{cp}K_{plli}}{3T_sT_{pll}} & \frac{8I_{ggd}K_{cp}K_{pllp}}{3T_sT_{pll}} & 0 & 0 & 0 \\
0 & 0 & 0 & 0 & 0 & 0 & 0 & 0 & 0 & 0 & 0 & 1 & 0 & 0 & 0 \\
0 & 0 & 0 & 0 & 0 & 0 & 0 & 0 & 0 & 0 & 0 & 0 & 1 & 0 & 0 \\
0 & 0 & 0 & \frac{L_s}{L_{gg}+L_s} & 0 & 0 & 0 & 0 & 0 & 0 & \frac{-V_{gd}K_{plli}}{T_{pll}} & \frac{-V_{gd}K_{pllp}}{T_{pll}} & \frac{-1}{T_{pll}} & 0 & 0 \\
0 & 0 & \frac{3I_{ggd}}{2T_p} & \frac{3I_{ggq}}{2T_p} & \frac{3V_{gfd}}{2T_p} & \frac{3V_{gfq}}{2T_p} & 0 & 0 & 0 & 0 & 0 & 0 & 0 & \frac{-1}{T_p} & 0 \\
0 & 0 & \frac{-3I_{ggq}}{2T_q} & \frac{3I_{ggd}}{2T_q} & \frac{3V_{gfq}}{2T_q} & \frac{-3V_{gfd}}{2T_q} & 0 & 0 & 0 & 0 & 0 & 0 & 0 & 0 & \frac{-1}{T_q}
\end{bmatrix}, \tag{4.52}
$$

$$
\mathbf{B}_{\mathbf{TypeI}} = \begin{bmatrix}
\frac{-K_{\mathrm{cp}}}{L_{\mathrm{gi}}} & 0 & 0 & 0 & 0 & 0 & 1 & 0 & \frac{8K_{\mathrm{cp}}}{3T_{\mathrm{s}}} & 0 & 0 & 0 & 0 & 0 & 0 \\
0 & \frac{-K_{\mathrm{cp}}}{L_{\mathrm{gi}}} & 0 & 0 & 0 & 0 & 0 & 1 & 0 & \frac{8K_{\mathrm{cp}}}{3T_{\mathrm{s}}} & 0 & 0 & 0 & 0 & 0 \\
0 & 0 & 0 & 0 & \frac{-1}{L_{\mathrm{gg}}+L_{\mathrm{s}}} & 0 & 0 & 0 & 0 & 0 & 0 & 0 & 0 & 0 & 0 \\
0 & 0 & 0 & 0 & 0 & \frac{-1}{L_{\mathrm{gg}}+L_{\mathrm{s}}} & 0 & 0 & 0 & 0 & 0 & 0 & \frac{L_{\mathrm{gg}}}{L_{\mathrm{gg}}+L_{\mathrm{s}}} & 0 & 0
\end{bmatrix}^{\mathrm{T}}, \text{ and} \tag{4.53}
$$

$$
\mathbf{C}_{\mathbf{TypeI}} = \begin{bmatrix}
0&0&0&0&1&0&0&0&0&0&0&0&0&0&0&0 \\
0&0&0&0&0&1&0&0&0&0&0&0&0&0&0&0 \\
0&0&0&0&0&0&0&0&0&0&0&0&\frac{K_{\mathrm{plli}}}{T_{\mathrm{pll}}}&\frac{K_{\mathrm{pllp}}}{T_{\mathrm{pll}}}&0&0 \\
0&0&0&0&0&0&0&0&0&0&0&0&0&0&1&0 \\
0&0&0&0&0&0&0&0&0&0&0&0&0&0&0&1
\end{bmatrix}. \tag{4.54}
$$

Next, Fig. 4.10 depicts the block representation of Type II GTCs. Remember that Type II GTCs control the dc voltage through ac currents. Moreover, PLLs are necessary in Type II GTCs for grid synchronization.

We derive the model of Type II GTCs from Fig. 4.10 as

$$
\begin{aligned}
\tfrac{\mathrm{d}}{\mathrm{d}t}\Delta\mathbf{x}_{\mathbf{TypeII}} &= \mathbf{A}_{\mathbf{TypeII}}\Delta\mathbf{x}_{\mathbf{TypeII}} + \mathrm{B}_{\mathbf{TypeII}}\Delta\mathbf{u}_{\mathbf{TypeII}} \\
\Delta\mathbf{y}_{\mathbf{TypeII}} &= \mathbf{C}_{\mathbf{TypeII}}\Delta\mathbf{x}_{\mathbf{TypeII}}
\end{aligned}, \tag{4.55}
$$

where

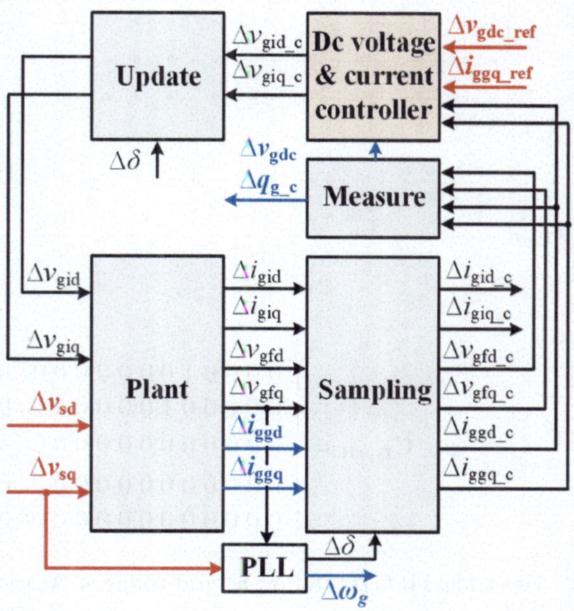

Fig. 4.10 Block representation of Type II GTCs

$$
\begin{aligned}
\Delta\mathbf{x}_{\mathbf{TypeII}(1\times 16)} &= \begin{bmatrix} \Delta\mathbf{x}_{\mathbf{plant}} & \Delta\mathbf{x}_{\mathbf{dc\_voltage}} & \Delta\mathbf{x}_{\mathbf{grid\_current}} & \Delta\mathbf{x}_{\mathbf{pll}} & \Delta\mathbf{x}_{\mathbf{vdcq\_measure}} \end{bmatrix}^{\mathrm{T}} \\
\Delta\mathbf{u}_{\mathbf{TypeII}} &= \begin{bmatrix} \Delta v_{\mathrm{gdc\_ref}} & \Delta i_{\mathrm{ggq\_ref}} & \Delta v_{\mathrm{sd}} & \Delta v_{\mathrm{sq}} \end{bmatrix}^{\mathrm{T}} \\
\Delta\mathbf{y}_{\mathbf{TypeII}} &= \begin{bmatrix} \Delta i_{\mathrm{ggd}} & \Delta i_{\mathrm{ggq}} & \Delta\omega_{\mathrm{g}} & \Delta v_{\mathrm{gdc}} & \Delta q_{\mathrm{g\_c}} \end{bmatrix}^{\mathrm{T}}
\end{aligned}
\quad . \qquad (4.56)
$$

As compared with Type I GTCs, Type II GTCs should regulate the dc voltage Δv_{gdc} on top of ac currents. Due to dc voltage regulation, the active power signal is removed from the output signals, while the remaining input and output signals are unchanged.

The matrices related to Type II GTCs can be derived from Fig. 4.10 as

$$
\mathbf{A}_{\mathbf{TypeII}} = \begin{bmatrix}
0 & \omega_0 & \frac{-1}{L_{gi}} & 0 & \frac{K_{cp}}{L_{gi}} & 0 & \frac{-K_{dci}K_{cp}}{L_{gi}} & \frac{-K_{ci}}{L_{gi}} & 0 & \frac{1}{L_{gi}} & 0 & \frac{K_{cp}K_{plli}I_{ggq}-V_{giq}K_{plli}}{T_{pll}L_{gi}} & \frac{K_{cp}K_{pllp}I_{ggq}-V_{giq}K_{pllp}}{T_{pll}L_{gi}} & 0 & \frac{-K_{dcp}K_{cp}}{L_{gi}} & 0 \\
-\omega_0 & 0 & 0 & \frac{-1}{L_{gi}} & 0 & \frac{K_{cp}}{L_{gi}} & 0 & 0 & \frac{-K_{ci}}{L_{gi}} & 0 & \frac{1}{L_{gi}} & \frac{-K_{cp}K_{plli}I_{ggd}+V_{gid}K_{plli}}{T_{pll}L_{gi}} & \frac{-K_{cp}K_{pllp}I_{ggd}+V_{gid}K_{pllp}}{T_{pll}L_{gi}} & 0 & 0 & 0 \\
\frac{1}{C_{gf}} & 0 & 0 & \omega_0 & \frac{-1}{C_{gf}} & 0 & 0 & 0 & 0 & 0 & 0 & 0 & 0 & 0 & 0 & 0 \\
0 & \frac{1}{C_{gf}} & -\omega_0 & 0 & 0 & \frac{-1}{C_{gf}} & 0 & 0 & 0 & 0 & 0 & 0 & 0 & 0 & 0 & 0 \\
0 & 0 & \frac{1}{L_{gg}+L_s} & 0 & 0 & \omega_0 & 0 & 0 & 0 & 0 & 0 & 0 & 0 & 0 & 0 & 0 \\
0 & 0 & 0 & \frac{1}{L_{gg}+L_s} & -\omega_0 & 0 & 0 & 0 & 0 & 0 & 0 & 0 & 0 & 0 & 0 & 0 \\
0 & 0 & 0 & 0 & 0 & 0 & 0 & 0 & 0 & 0 & 0 & 0 & 0 & 0 & 1 & 0 \\
0 & 0 & 0 & 0 & -1 & 0 & K_{dci} & 0 & 0 & 0 & 0 & \frac{-K_{plli}I_{ggq}}{T_{pll}} & \frac{-K_{pllp}I_{ggq}}{T_{pll}} & 0 & K_{dcp} & 0 \\
0 & 0 & 0 & 0 & 0 & -1 & 0 & 0 & 0 & 0 & 0 & \frac{K_{plli}I_{ggd}}{T_{pll}} & \frac{K_{pllp}I_{ggd}}{T_{pll}} & 0 & 0 & 0 \\
0 & 0 & 0 & 0 & \frac{-8K_{cp}}{3T_s} & 0 & \frac{8K_{dci}K_{cp}}{3T_s} & \frac{8K_{ci}}{3T_s} & 0 & \frac{-4}{3T_s} & 0 & \frac{-8K_{cp}K_{plli}I_{ggq}}{3T_sT_{pll}} & \frac{-8K_{cp}K_{pllp}I_{ggq}}{3T_sT_{pll}} & 0 & \frac{8K_{dcp}K_{cp}}{3T_s} & 0 \\
0 & 0 & 0 & 0 & 0 & \frac{-8K_{cp}}{3T_s} & 0 & 0 & \frac{8K_{ci}}{3T_s} & 0 & \frac{-4}{3T_s} & \frac{8K_{cp}K_{plli}I_{ggd}}{3T_sT_{pll}} & \frac{8K_{cp}K_{pllp}I_{ggd}}{3T_sT_{pll}} & 0 & 0 & 0 \\
0 & 0 & 0 & 0 & 0 & 0 & 0 & 0 & 0 & 0 & 0 & 0 & 1 & 0 & 0 & 0 \\
0 & 0 & 0 & 0 & 0 & 0 & 0 & 0 & 0 & 0 & 0 & 0 & 0 & 1 & 0 & 0 \\
0 & 0 & 0 & \frac{L_s}{L_{gg}+L_s} & 0 & 0 & 0 & 0 & 0 & 0 & 0 & \frac{-V_{gd}K_{plli}}{T_{pll}} & \frac{-V_{gd}K_{pllp}}{T_{pll}} & \frac{-1}{T_{pll}} & 0 & 0 \\
0 & 0 & \frac{-3I_{ggd}}{2C_{gdc}V_{gdc}} & \frac{-3I_{ggq}}{2C_{gdc}V_{gdc}} & \frac{-3V_{gfd}}{2C_{gdc}V_{gdc}} & \frac{-3V_{gfq}}{2C_{gdc}V_{gdc}} & 0 & 0 & 0 & 0 & 0 & 0 & 0 & 0 & 0 & 0 \\
0 & 0 & \frac{-3I_{ggq}}{2T_q} & \frac{3I_{ggd}}{2T_q} & \frac{3V_{gfq}}{2T_q} & \frac{-3V_{gfd}}{2T_q} & 0 & 0 & 0 & 0 & 0 & 0 & 0 & 0 & 0 & \frac{-1}{T_q}
\end{bmatrix}, \qquad (4.57)
$$

$$
\mathbf{B}_{\mathbf{TypeII}} = \begin{bmatrix}
\frac{K_{dcp}K_{cp}}{L_{gi}} & 0 & 0 & 0 & 0 & 0 & -1 & -K_{dcp} & 0 & \frac{-8K_{dcp}K_{cp}}{3T_s} & 0 & 0 & 0 & 0 & 0 & 0 \\
0 & \frac{-K_{cp}}{L_{gi}} & 0 & 0 & 0 & 0 & 0 & 0 & 1 & 0 & \frac{8K_{cp}}{3T_s} & 0 & 0 & 0 & 0 & 0 \\
0 & 0 & 0 & 0 & \frac{-1}{L_{gg}+L_s} & 0 & 0 & 0 & 0 & 0 & 0 & 0 & 0 & 0 & 0 & 0 \\
0 & 0 & 0 & 0 & 0 & \frac{-1}{L_{gg}+L_s} & 0 & 0 & 0 & 0 & 0 & 0 & 0 & \frac{L_{gg}}{L_{gg}+L_s} & 0 & 0
\end{bmatrix}^{\mathrm{T}}, \text{ and} \qquad (4.58)
$$

$$
\mathbf{C}_{\mathbf{TypeII}} = \begin{bmatrix}
0 & 0 & 0 & 0 & 1 & 0 & 0 & 0 & 0 & 0 & 0 & 0 & 0 & 0 & 0 & 0 \\
0 & 0 & 0 & 0 & 0 & 1 & 0 & 0 & 0 & 0 & 0 & 0 & 0 & 0 & 0 & 0 \\
0 & 0 & 0 & 0 & 0 & 0 & 0 & 0 & 0 & 0 & 0 & 0 & \frac{K_{plli}}{T_{pll}} & \frac{K_{pllp}}{T_{pll}} & 0 & 0 \\
0 & 0 & 0 & 0 & 0 & 0 & 0 & 0 & 0 & 0 & 0 & 0 & 0 & 0 & 1 & 0 \\
0 & 0 & 0 & 0 & 0 & 0 & 0 & 0 & 0 & 0 & 0 & 0 & 0 & 0 & 0 & 1
\end{bmatrix}. \qquad (4.59)
$$

Types I and II GTCs follow ac grid voltages. As such, they necessitate PLLs and ac grid current controllers. In comparison, Type I GTCs regulate active power due to their fixed dc buses. Instead, Type II GTCs control dc voltages, thus enabling dc

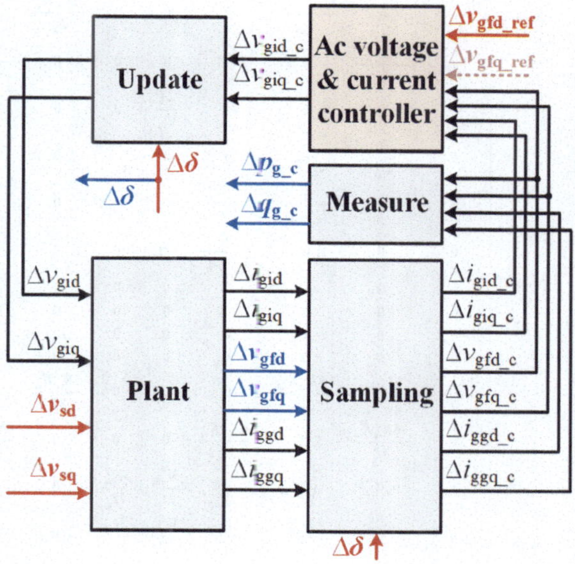

Fig. 4.11 Block representation of Type III GTCs

grid formulation. The overall models derived above allow us to analyze the stability of Types I and II GTCs.

We continue to the case of Type III GTCs. For visualization, Fig. 4.11 presents the relevant block representation, where PLLs are excluded, as ac grid-forming GTCs use external phase angle references in replacement of PLL phase angles. To form ac grids, Type III GTCs should regulate ac voltages Δv_{gfd} and Δv_{gfq} to their references Δv_{gfd_ref} and Δv_{gfq_ref}, respectively. Note that Δv_{gfq_ref} is redundant, since it is determined by the phase angle $\Delta\delta$ and Δv_{gfd_ref}.

We build the overall model of Type III GTCs as

$$\begin{aligned} \frac{d}{dt}\Delta \mathbf{x}_{\mathbf{TypeIII}} &= \mathbf{A}_{\mathbf{TypeIII}}\Delta \mathbf{x}_{\mathbf{TypeIII}} + \mathrm{B}_{\mathbf{TypeIII}}\Delta \mathbf{u}_{\mathbf{TypeIII}} \\ \Delta \mathbf{y}_{\mathbf{TypeIII}} &= \mathbf{C}_{\mathbf{TypeIII}}\Delta \mathbf{x}_{\mathbf{TypeIII}} + \mathbf{D}_{\mathbf{TypeIII}}\Delta \mathbf{u}_{\mathbf{TypeIII}} \end{aligned}, \tag{4.60}$$

where

$$\begin{aligned} \Delta \mathbf{x}_{\mathbf{TypeIII}(1\times 14)} &= \left[\, \Delta \mathbf{x}_{\mathbf{plant}}\ \Delta \mathbf{x}_{\mathbf{ac\_voltage}}\ \Delta \mathbf{x}_{\mathbf{con\_current}}\ \Delta \mathbf{x}_{\mathbf{pq\_measure}} \,\right]^{\mathrm{T}} \\ \Delta \mathbf{u}_{\mathbf{TypeIII}} &= \left[\, \Delta v_{gfd\_ref}\ \Delta v_{gfq\_ref}\ \Delta\delta\ \Delta v_{sd}\ \Delta v_{sq} \,\right]^{\mathrm{T}} \\ \Delta \mathbf{y}_{\mathbf{TypeIII}} &= \left[\, \Delta v_{gfd}\ \Delta v_{gfq}\ \Delta\delta\ \Delta p_{g\_c}\ \Delta q_{g\_c} \,\right]^{\mathrm{T}} \end{aligned}. \tag{4.61}$$

The input signals contain ac voltage (Δv_{gfd_ref} and Δv_{gfq_ref}) and phase angle ($\Delta\delta$) references, while the output signals comprise ac voltages (Δv_{gfd} and Δv_{gfq}). In addition, the phase angle ($\Delta\delta$), active power (Δp_{g_c}), and reactive power (i.e., Δq_{g_c}) are maintained as outputs for outer control loops, which aim to provide grid-supportive services, as will be detailed.

We derive the corresponding matrices as

$$\mathbf{A}_{\mathbf{TypeIII}} = \begin{bmatrix} \frac{K_{cp}K_{id}}{L_{gi}} & \omega_0 & \frac{K_{cp}K_{vp}-1}{L_{gi}} & 0 & 0 & 0 & \frac{-K_{cp}K_{vi}}{L_{gi}} & 0 & \frac{-K_{ci}}{L_{gi}} & 0 & \frac{1}{L_{gi}} & 0 & 0 & 0 \\ -\omega_0 & \frac{K_{cp}K_{id}}{L_{gi}} & 0 & \frac{K_{cp}K_{vp}-1}{L_{gi}} & 0 & 0 & 0 & \frac{-K_{cp}K_{vi}}{L_{gi}} & 0 & \frac{-K_{ci}}{L_{gi}} & 0 & \frac{1}{L_{gi}} & 0 & 0 \\ \frac{1}{C_{gf}} & 0 & 0 & \omega_0 & \frac{-1}{C_{gf}} & 0 & 0 & 0 & 0 & 0 & 0 & 0 & 0 & 0 \\ 0 & \frac{1}{C_{gf}} & -\omega_0 & 0 & 0 & \frac{-1}{C_{gf}} & 0 & 0 & 0 & 0 & 0 & 0 & 0 & 0 \\ 0 & 0 & \frac{1}{L_{gg}+L_s} & 0 & 0 & \omega_0 & 0 & 0 & 0 & 0 & 0 & 0 & 0 & 0 \\ 0 & 0 & 0 & \frac{1}{L_{gg}+L_s} & -\omega_0 & 0 & 0 & 0 & 0 & 0 & 0 & 0 & 0 & 0 \\ 0 & 0 & -1 & 0 & 0 & 0 & 0 & 0 & 0 & 0 & 0 & 0 & 0 & 0 \\ 0 & 0 & 0 & -1 & 0 & 0 & 0 & 0 & 0 & 0 & 0 & 0 & 0 & 0 \\ -K_{id} & 0 & -K_{vp} & 0 & 0 & 0 & K_{vi} & 0 & 0 & 0 & 0 & 0 & 0 & 0 \\ 0 & -K_{id} & 0 & -K_{vp} & 0 & 0 & 0 & K_{vi} & 0 & 0 & 0 & 0 & 0 & 0 \\ \frac{-8K_{cp}K_{id}}{3T_s} & 0 & \frac{-8K_{cp}K_{vp}}{3T_s} & 0 & 0 & 0 & \frac{8K_{cp}K_{vi}}{3T_s} & 0 & \frac{8K_{ci}}{3T_s} & 0 & \frac{-4}{3T_s} & 0 & 0 & 0 \\ 0 & \frac{-8K_{cp}K_{id}}{3T_s} & 0 & \frac{-8K_{cp}K_{vp}}{3T_s} & 0 & 0 & 0 & \frac{8K_{cp}K_{vi}}{3T_s} & 0 & \frac{8K_{ci}}{3T_s} & 0 & \frac{-4}{3T_s} & 0 & 0 \\ 0 & 0 & \frac{3I_{ggd}}{2T_p} & \frac{3I_{ggq}}{2T_p} & \frac{3V_{gfd}}{2T_p} & \frac{3V_{gfq}}{2T_p} & 0 & 0 & 0 & 0 & 0 & 0 & \frac{-1}{T_p} & 0 \\ 0 & 0 & \frac{-3I_{ggq}}{2T_q} & \frac{3I_{ggd}}{2T_q} & \frac{3V_{gfq}}{2T_q} & \frac{-3V_{gfd}}{2T_q} & 0 & 0 & 0 & 0 & 0 & 0 & 0 & \frac{-1}{T_q} \end{bmatrix}, \tag{4.62}$$

$$\mathbf{B}_{\mathbf{TypeIII}} = \begin{bmatrix} \frac{-K_{cp}K_{vp}}{L_{gi}} & 0 & \frac{I_{giq}K_{cp}K_{id}+V_{gfq}K_{cp}K_{vp}-V_{giq}}{L_{gi}} & 0 & 0 \\ 0 & \frac{-K_{cp}K_{vp}}{L_{gi}} & \frac{-I_{gid}K_{cp}K_{id}-V_{gfd}K_{cp}K_{vp}+V_{gid}}{L_{gi}} & 0 & 0 \\ 0 & 0 & 0 & 0 & 0 \\ 0 & 0 & 0 & 0 & 0 \\ 0 & 0 & 0 & \frac{-1}{L_{gg}+L_s} & 0 \\ 0 & 0 & 0 & 0 & \frac{-1}{L_{gg}+L_s} \\ 1 & 0 & -V_{gfq} & 0 & 0 \\ 0 & 1 & V_{gfd} & 0 & 0 \\ K_{vp} & 0 & -I_{giq}K_{id} - V_{gfq}K_{vp} & 0 & 0 \\ 0 & K_{vp} & I_{gid}K_{id} + V_{gfd}K_{vp} & 0 & 0 \\ \frac{8K_{cp}K_{vp}}{3T_s} & 0 & \frac{-8I_{giq}K_{cp}K_{id}-8V_{gfq}K_{cp}K_{vp}}{3T_s} & 0 & 0 \\ 0 & \frac{8K_{cp}K_{vp}}{3T_s} & \frac{8I_{gid}K_{cp}K_{id}+8V_{gfd}K_{cp}K_{vp}}{3T_s} & 0 & 0 \\ 0 & 0 & 0 & 0 & 0 \\ 0 & 0 & 0 & 0 & 0 \end{bmatrix}, \tag{4.63}$$

$$\mathbf{C}_{\mathbf{TypeIII}} = \begin{bmatrix} 0&0&1&0&0&0&0&0&0&0&0&0&0&0 \\ 0&0&0&1&0&0&0&0&0&0&0&0&0&0 \\ 0&0&0&0&0&0&0&0&0&0&0&0&0&0 \\ 0&0&0&0&0&0&0&0&0&0&0&0&1&0 \\ 0&0&0&0&0&0&0&0&0&0&0&0&0&1 \end{bmatrix}, \text{ and} \tag{4.64}$$

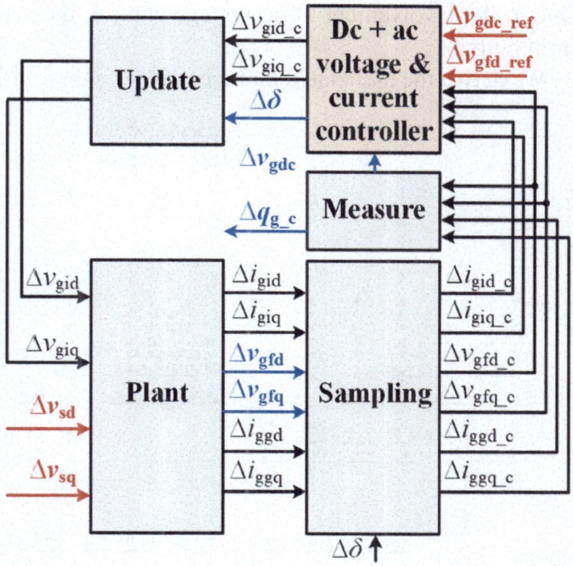

Fig. 4.12 Block representation of Type IV GTCs

$$\mathbf{D}_{\mathbf{TypeIII}} = \begin{bmatrix} 0\,0\,0\,0\,0 \\ 0\,0\,0\,0\,0 \\ 0\,0\,1\,0\,0 \\ 0\,0\,0\,0\,0 \\ 0\,0\,0\,0\,0 \end{bmatrix}. \tag{4.65}$$

Finally, Fig. 4.12 demonstrates the block representation of Type IV GTCs, which provide grid formulation in both ac and dc sides. Owning to the ac grid-forming feature, PLLs are removed. Because of the dc grid-forming capability, the dc voltage v_{gdc} is dedicatedly controlled.

We model Type IV GTCs through

$$\begin{aligned} \tfrac{\mathrm{d}}{\mathrm{d}t}\Delta\mathbf{x}_{\mathbf{TypeIV}} &= \mathbf{A}_{\mathbf{TypeIV}}\Delta\mathbf{x}_{\mathbf{TypeIV}} + \mathrm{B}_{\mathbf{TypeIV}}\Delta\mathbf{u}_{\mathbf{TypeIV}} \\ \Delta\mathbf{y}_{\mathbf{TypeIV}} &= \mathbf{C}_{\mathbf{TypeIV}}\Delta\mathbf{x}_{\mathbf{TypeIV}} + \mathbf{D}_{\mathbf{TypeIV}}\Delta\mathbf{u}_{\mathbf{TypeIV}} \end{aligned}, \tag{4.66}$$

where

$$\begin{aligned} \Delta\mathbf{x}_{\mathbf{TypeIV}(1\times15)} &= \left[\, \Delta\mathbf{x}_{\mathbf{plant}}\ \Delta\mathbf{x}_{\mathbf{dc\_voltage}}\ \Delta\mathbf{x}_{\mathbf{ac\_voltage}}\ \Delta\mathbf{x}_{\mathbf{con\_current}}\ \Delta\mathbf{x}_{\mathbf{vdcq\_measure}} \,\right]^{\mathrm{T}} \\ \Delta\mathbf{u}_{\mathbf{TypeIV}} &= \left[\, \Delta v_{\mathrm{gdc\_ref}}\ \Delta v_{\mathrm{gfd\_ref}}\ \Delta v_{\mathrm{gfq\_ref}}\ \Delta v_{\mathrm{sd}}\ \Delta v_{\mathrm{sq}} \,\right]^{\mathrm{T}} \\ \Delta\mathbf{y}_{\mathbf{TypeIV}} &= \left[\, \Delta v_{\mathrm{gfd}}\ \Delta v_{\mathrm{gfq}}\ \Delta\delta\ \Delta v_{\mathrm{gdc}}\ \Delta q_{\mathrm{g\_c}} \,\right]^{\mathrm{T}} \end{aligned}. \tag{4.67}$$

As compared with Type III GTCs, Type IV GTCs treat the phase angle reference as an internal reference given by the dc voltage controller instead of an input signal.

Due to the involvement of dc voltage control, the active power signal is excluded from outputs.

We derive the associated matrices as

$$\mathbf{A}_{\mathbf{TypeIV}}=\left[\begin{array}{ccccccccccccccc}
\frac{K_{cp}K_{id}}{L_{gi}} & \omega_b & \frac{K_{cp}K_{vp}-1}{L_{gi}} & 0 & 0 & 0 & \frac{-V_{giq}K_{dci}+I_{giq}K_{dci}K_{cp}K_{id}+V_{gfq}K_{dci}K_{cp}K_{vp}}{L_{gi}} & \frac{-K_{cp}K_{vi}}{L_{gi}} & 0 & \frac{-K_{ci}}{L_{gi}} & 0 & \frac{1}{L_{gi}} & 0 & \frac{-V_{giq}K_{dcp}+I_{giq}K_{dcp}K_{cp}K_{id}+V_{gfq}K_{dcp}K_{cp}K_{vp}}{L_{gi}} & 0\\
-\omega_b & \frac{K_{cp}K_{id}}{L_{gi}} & 0 & \frac{K_{cp}K_{vp}-1}{L_{gi}} & 0 & 0 & \frac{V_{gid}K_{dci}-I_{gid}K_{dci}K_{cp}K_{id}-V_{gfd}K_{dci}K_{cp}K_{vp}}{L_{gi}} & 0 & \frac{-K_{cp}K_{vi}}{L_{gi}} & 0 & -K_{ci} & 0 & \frac{1}{L_{gi}} & \frac{V_{gid}K_{dcp}-I_{gid}K_{dcp}K_{cp}K_{id}-V_{gfd}K_{dcp}K_{cp}K_{vp}}{L_{gi}} & 0\\
\frac{1}{C_{gf}} & 0 & 0 & \omega_b & \frac{-1}{C_{gf}} & 0 & 0 & 0 & 0 & 0 & 0 & 0 & 0 & 0 & 0\\
0 & \frac{1}{C_{gf}} & -\omega_b & 0 & 0 & \frac{-1}{C_{gf}} & 0 & 0 & 0 & 0 & 0 & 0 & 0 & 0 & 0\\
0 & 0 & \frac{1}{L_{gg}+L_s} & 0 & 0 & \omega_b & 0 & 0 & 0 & 0 & 0 & 0 & 0 & 0 & 0\\
0 & 0 & 0 & \frac{1}{L_{gg}+L_s} & -\omega_b & 0 & 0 & 0 & 0 & 0 & 0 & 0 & 0 & 0 & 0\\
0 & 0 & 0 & 0 & 0 & 0 & 0 & 0 & 0 & 0 & 0 & 0 & 0 & 1 & 0\\
0 & 0 & -1 & 0 & 0 & 0 & -V_{gfq}K_{dci} & 0 & 0 & 0 & 0 & 0 & 0 & -V_{gfq}K_{dcp} & 0\\
0 & 0 & 0 & -1 & 0 & 0 & V_{gfd}K_{dci} & 0 & 0 & 0 & 0 & 0 & 0 & V_{gfd}K_{dcp} & 0\\
-K_{id} & 0 & -K_{vp} & 0 & 0 & 0 & -I_{giq}K_{dci}K_{id}-V_{gfq}K_{dci}K_{vp} & K_{vi} & 0 & 0 & 0 & 0 & 0 & -I_{giq}K_{dcp}K_{id}-V_{gfq}K_{dcp}K_{vp} & 0\\
0 & -K_{id} & 0 & -K_{vp} & 0 & 0 & I_{gid}K_{dci}K_{id}+V_{gfd}K_{dci}K_{vp} & 0 & K_{vi} & 0 & 0 & 0 & 0 & I_{gid}K_{dcp}K_{id}+V_{gfd}K_{dcp}K_{vp} & 0\\
\frac{-8K_{cp}K_{id}}{3T_s} & 0 & \frac{-8K_{cp}K_{vp}}{3T_s} & 0 & 0 & 0 & \frac{-8I_{giq}K_{dci}K_{cp}K_{id}-8V_{gfq}K_{dci}K_{cp}K_{vp}}{3T_s} & \frac{8K_{cp}K_{vi}}{3T_s} & 0 & \frac{8K_{ci}}{3T_s} & 0 & \frac{-4}{3T_s} & 0 & \frac{-8I_{giq}K_{dcp}K_{cp}K_{id}-8V_{gfq}K_{dcp}K_{cp}K_{vp}}{3T_s} & 0\\
0 & \frac{-8K_{cp}K_{id}}{3T_s} & 0 & \frac{-8K_{cp}K_{vp}}{3T_s} & 0 & 0 & \frac{8I_{gid}K_{dci}K_{cp}K_{id}+8V_{gfd}K_{dci}K_{cp}K_{vp}}{3T_s} & 0 & \frac{8K_{cp}K_{vi}}{3T_s} & 0 & \frac{8K_{ci}}{3T_s} & 0 & \frac{-4}{3T_s} & \frac{8I_{gid}K_{dcp}K_{cp}K_{id}+8V_{gfd}K_{dcp}K_{cp}K_{vp}}{3T_s} & 0\\
0 & 0 & \frac{-3I_{gid}}{2C_{gdc}V_{gdc}} & \frac{-3I_{giq}}{2C_{gdc}V_{gdc}} & \frac{-3V_{gfd}}{2C_{gdc}V_{gdc}} & \frac{-3V_{gfq}}{2C_{gdc}V_{gdc}} & 0 & 0 & 0 & 0 & 0 & 0 & 0 & 0 & 0\\
0 & 0 & \frac{-3I_{giq}}{2T_q} & \frac{3I_{gid}}{2T_q} & \frac{3V_{gfq}}{2T_q} & \frac{-3V_{gfd}}{2T_q} & 0 & 0 & 0 & 0 & 0 & 0 & 0 & 0 & \frac{-1}{T_q}
\end{array}\right], \tag{4.68}$$

$$\mathbf{B}_{\mathbf{TypeIV}}=\left[\begin{array}{ccccc}
\frac{V_{giq}K_{dcp}-I_{giq}K_{dcp}K_{cp}K_{id}-V_{gfq}K_{dcp}K_{cp}K_{vp}}{L_{gi}} & \frac{-K_{cp}K_{vp}}{L_{gi}} & 0 & 0 & 0\\
\frac{-V_{gid}K_{dcp}+I_{gid}K_{dcp}K_{cp}K_{id}+V_{gfd}K_{dcp}K_{cp}K_{vp}}{L_{gi}} & 0 & \frac{-K_{cp}K_{vp}}{L_{gi}} & 0 & 0\\
0 & 0 & 0 & 0 & 0\\
0 & 0 & 0 & 0 & 0\\
0 & 0 & 0 & \frac{-1}{L_{gg}+L_s} & 0\\
0 & 0 & 0 & 0 & \frac{-1}{L_{gg}+L_s}\\
-1 & 0 & 0 & 0 & 0\\
V_{gfq}K_{dcp} & 1 & 0 & 0 & 0\\
-V_{gfd}K_{dcp} & 0 & 1 & 0 & 0\\
I_{giq}K_{dcp}K_{id}+V_{gfq}K_{dcp}K_{vp} & K_{vp} & 0 & 0 & 0\\
-I_{gid}K_{dcp}K_{id}-V_{gfd}K_{dcp}K_{vp} & 0 & K_{vp} & 0 & 0\\
\frac{8I_{giq}K_{dcp}K_{cp}K_{id}+8V_{gfq}K_{dcp}K_{cp}K_{vp}}{3T_s} & \frac{8K_{cp}K_{vp}}{3T_s} & 0 & 0 & 0\\
\frac{-8I_{gid}K_{dcp}K_{cp}K_{id}-8V_{gfd}K_{dcp}K_{cp}K_{vp}}{3T_s} & 0 & \frac{8K_{cp}K_{vp}}{3T_s} & 0 & 0\\
0 & 0 & 0 & 0 & 0\\
0 & 0 & 0 & 0 & 0
\end{array}\right], \tag{4.69}$$

$$\mathbf{C}_{\mathbf{TypeIV}}=\left[\begin{array}{ccccccccccccccc}
0&0&1&0&0&0&0&0&0&0&0&0&0&0&0\\
0&0&0&1&0&0&0&0&0&0&0&0&0&0&0\\
0&0&0&0&0&0&K_{dci}&0&0&0&0&0&0&K_{dcp}&0\\
0&0&0&0&0&0&0&0&0&0&0&0&0&1&0\\
0&0&0&0&0&0&0&0&0&0&0&0&0&0&1
\end{array}\right],\ \text{and} \tag{4.70}$$

$$\mathbf{D}_{\mathbf{TypeIV}} = \begin{bmatrix} 0 & 0\,0\,0\,0 \\ 0 & 0\,0\,0\,0 \\ -K_{\mathrm{dcp}} & 0\,0\,0\,0 \\ 0 & 0\,0\,0\,0 \\ 0 & 0\,0\,0\,0 \end{bmatrix}. \tag{4.71}$$

Types III and IV GTCs form ac grid voltages. As a result, they exclude PLLs and provide external phase angles. The above overall models enable the stability analyses of Types III and IV GTCs.

We have finished the state-space modeling of the four types of GTCs, which covers the modeling of individual controllers, measurements, and other parts. Now, we are ready to investigate stability problems in more-electronics power systems. Let us start with converter-level stability problems.

4.2 Converter-Level Stability Problems

Converter-level stability problems cause instability of power converters. In more-electronics power systems, converter-level stability problems may easily lead to system-level stability problems and blackouts if not properly treated. Therefore, we must take care of converter-level stability problems.

In this section, we classify converter-level stability problems according to their sources or mechanisms. Specifically, we divide the reasons for converter-level stability problems into improper controller designs, time delays, weak grids, and poor power quality. Such a classification helps to better grasp the nature of stability problems as well as their solutions. Alternatively, stability problems can be classified according to time scales, which will also be used in the subsections wherever appropriate.

As mentioned, there are two methodologies (namely, block-diagram and state-space methodologies) for stability analysis. For clarity, we will choose a relatively simple and clear methodology in treating a specific problem. In addition, it should be remembered that the most straight forward approach of stability evaluation is through closed-loop poles in the continuous or discrete domain.

4.2.1 Improper Controller Designs

Improper controller designs are the most common inducement of instability. Notably, improper controllers refer to those controllers used under standard operating conditions of power converters, where power grids are stiff grids with high power quality. In this case, designers should be responsible for unstable controllers.

4.2.1.1 Ac Current Controllers

As mentioned in Sect. 4.1, we employ ac current controllers in all four types of GTCs. Therefore, ac current controllers are of paramount significance and should be designed first.

Generally, we use PI regulators as ac current controllers. As discussed in Sect. 3.4, the integrator often plays a minor role on system stability (see Fig. 3.37). Instead, the proportional gain should carefully be selected. Referring to Figs. 3.35 and 3.36, we note that proportional gains must be maintained within their maximum allowable ranges. Otherwise, instability occur. Therefore, an improper current controller often features an excessively large current proportional gain.

To illustrate the effect of improper current controllers, we design Type I GTCs according to Table 4.2, where two current control proportional gains (i.e., $K_{cp} = 3$ and 30) are involved. Figure 4.13 shows the pole-zero maps of Type I GTCs based on the state-space models built in Sect. 4.1. Clearly, the increment of K_{cp} makes system unstable, as can be validated from its right-half-plane poles.

Pole-zero maps visualize the trend of pole position changes. However, we may not easily identify unstable poles when the number of poles becomes large, particularly for well-separated poles. To fix this issue, we can calculate the numerical solutions of poles (or the eigenvalues of **A** matrix) and then compare their real parts with zero. The poles whose real parts are greater than zero are essentially unstable poles. Given that all the poles are with negative real parts, the system is stable. In Fig. 4.13,

Table 4.2 Type I GTC system and control parameters

Descriptions	Symbols	Values
Filter capacitance	C_{gf}	3 μF
Grid filter inductance	L_{gg}	5 mH
Converter filter inductance	L_{gi}	2 mH
Dc-link voltage reference	V_{gdc_ref}	400 V
Grid voltage reference (rms)	V_{s_ref}	110 V
Current reference in the *d*-axis	I_{ggd_ref}	10 A
Current reference in the *q*-axis	I_{ggq_ref}	0 A
Fundamental frequency	f_o	50 Hz
Resonance frequency	f_{cr}	2.43 kHz
Switching/sampling frequency	f_{sw}/f_s	10 kHz
Current control P gain	$\mathbf{K_{cp}}$	**3/30**
Current control I gain	K_{ci}	100
PLL control P gain	K_{pllp}	3
PLL control I gain	K_{plli}	300
PLL filter time constant	T_{pll}	$1/(2\pi \times 500)$

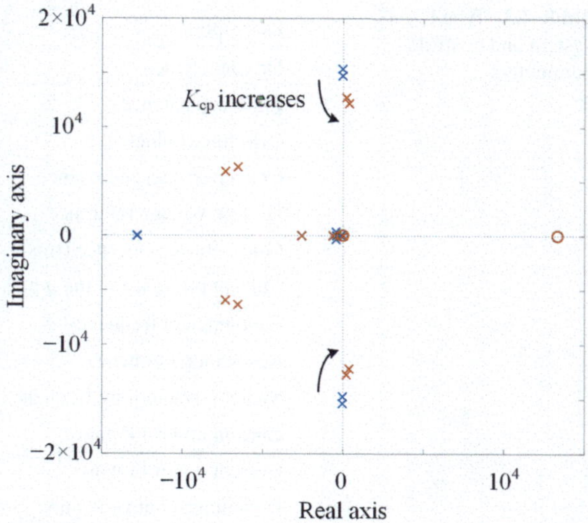

Fig. 4.13 Pole-zero maps of Type I GTCs as a function of current control P gain K_{cp}

as K_{cp} increases, the system changes from stable to unstable due to four unstable poles—195.91 ± 12,733.46*j* and 364.33 ± 12,192.82*j*.

Filter Resonances

Passive filter resonances pose an extra burden on the design of ac current controllers. Taking *LCL* filters as an example, we should design current controllers according to the resonance frequency f_{cr} of *LCL* filters. The specific allowable range of f_{cr} is closely related to time delays (e.g., introduced by reference calculation and PWM generation). In the case of one sampling period delay (where the sampling period equals the switching period) and ZOH, the converter-current-feedback controller (respectively, the grid-current-feedback controller) should be used when $f_r < f_s/6$ (respectively, $f_r > f_s/6$).

4.2.1.2 Dc Voltage Controllers

Dc voltage controllers are required in dc grid-forming GTCs, namely, Types II and IV GTCs. Given that ac current controllers are well tuned, dc voltage controllers may cause instability if not properly designed. Normally, we employ PI regulators as dc voltage controllers.

Next, we design Type II GTCs with the parameters listed in Table 4.3. As compared with Table 4.2, Table 4.3 further includes dc capacitance C_{gdc}, dc voltage control P gain K_{dcp}, and dc voltage control I gain K_{dci}. Moreover, the *d*-axis current reference is removed, as it acts as an internal variable now.

Figure 4.14 presents the pole-zero maps of Type II GTCs as a function of voltage control I gain K_{dci}. As observed from the zoom-in inset, an excessive value of the dc voltage control I gain K_{dci} introduces a pair of unstable poles (2.58 ± 397.46*j*),

Table 4.3 Type II GTC system and control parameters

Descriptions	Symbols	Values
Dc capacitance	C_{gdc}	2.82 mF
Filter capacitance	C_{gf}	3 μF
Grid filter inductance	L_{gg}	5 mH
Converter filter inductance	L_{gi}	2 mH
Dc-link voltage reference	V_{gdc_ref}	400 V
Grid voltage reference (rms)	V_{s_ref}	110 V
Current reference in the *q*-axis	I_{ggq_ref}	0 A
Fundamental frequency	f_o	50 Hz
Resonance frequency	f_{cr}	2.43 kHz
Switching/sampling frequency	f_{sw}/f_s	10 kHz
Current control P gain	K_{cp}	3
Current control I gain	K_{ci}	100
Dc voltage control P gain	K_{dcp}	1
Dc voltage control I gain	$\boldsymbol{K_{dci}}$	**10/1000**
PLL control P gain	K_{pllp}	3
PLL control I gain	K_{plli}	300
PLL filter time constant	T_{pll}	1/(2π × 500)

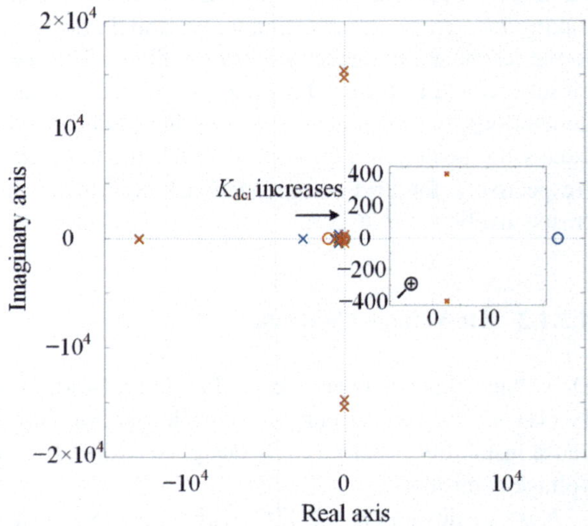

Fig. 4.14 Pole-zero maps of Type II GTCs as a function of voltage control I gain K_{dci}

leading to instability. Therefore, improper voltage control I gains are sources of instability. Although we change I gains in Fig. 4.14, it should be kept in mind that improper voltage control P gains may also introduce stability problems, which is similar to the case shown in Fig. 4.13.

It should be remined that the prerequisite of stable dc voltage control is stable ac current control. However, the opposite does not hold valid. When designing Type II GTCs, we first tune current controllers and then design dc voltage controllers. To decouple current and voltage controllers, they are often designed with well separated control bandwidths. Specifically, we design current controllers with wider control bandwidths and faster dynamics, thereby facilitating the design of dc voltage controllers. In terms of stability, these controllers should both be stabilized to ensure an overall stable system. Through observation of the locations of poles, we may possibly classify them according to various controllers, which helps individual controller tuning.

4.2.1.3 PLLs

We employ PLLs in ac grid-following GTCs, including Types I and II GTCs. PLLs aim to track the phase angle of grid voltages. Since PLLs introduce additional control blocks and transfer functions, they may also yield stability problems.

Once again, we take Type II GTCs as an example, where the parameters are documented in Table 4.3. In this case, we adopt a stable dc voltage I gain K_{dci} of 10. Meanwhile, two PLL filter time constants T_{pllp} of $1/(2\pi \times 500)$ and $1/(2\pi \times 10)$ are separately used. Figure 4.15 shows the relevant pole-zero maps of Type II GTCs. Obviously, an excessively large PLL time constant introduces unstable poles. This implies a tradeoff between harmonic filtering and stability of PLLs. Notably, we can use alternative PLL filters with higher orders, leading to the simultaneous improvements of filtering and stability. In addition to filter time constants, PLL control P and I gains should also properly be designed to avoid stability problems.

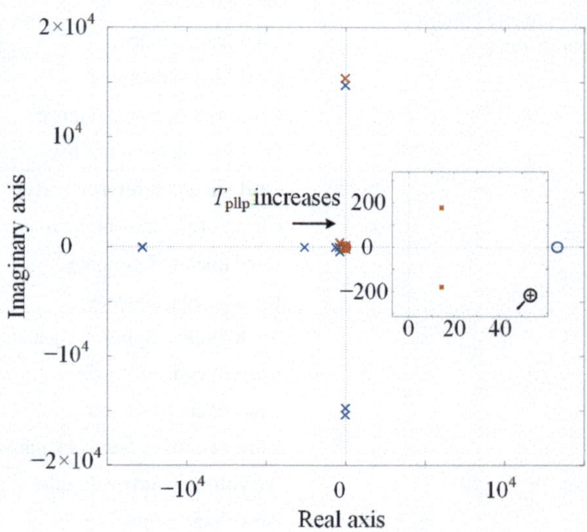

Fig. 4.15 Pole-zero maps of Type II GTCs as a function of PLL time constants T_{pllp}

Generally, PLLs themselves are well tuned offline. Therefore, they seldom introduce stability problems under normal operating conditions. However, in weak grids, PLLs are often instability sources. We will detail this point in the later sections. It is worth mentioning that PLLs are not the only choice for grid synchronization. A notable alternative option is the frequency-locked loop (FLL), which tracks the grid frequency instead of the grid phase angle [6]. However, PLLs are still dominant in practical applications.

4.2.1.4 Ac Voltage Controllers

We employ ac voltage controllers in ac grid-forming GTCs (i.e., Types III and IV GTCs). With properly designed current controllers, improper ac voltage controllers can cause stability problems. In Sect. 3.2, we have detailed the design procedure of ac voltage controllers, which comprise a PI regulator in combination with a repetitive controller. Ignoring the repetitive controller, we mainly focus on PI voltage regulators in this part.

We design Type III GTCs according to the parameters listed in Table 4.4. Due to the involvement of converter-current feedback control, the resonance frequency of *LCL* filters f_{cr} should be lower than $f_s/6$ to ensure stable systems. According to Table 4.4, we choose a f_{cr} of 665.79 Hz, which satisfies $f_{cr} < f_s/6$. This requires a relatively large filter capacitance of 40 μF.

Following the design procedure of UPS controllers in Sect. 3.2, we design the ac current controller as a P controller with a unity P gain (i.e., $K_{cp} = 1$) and a feedback gain K_{id} of 4. For simplification, we employ PI regulators (rather than PI

Table 4.4 Type III GTC system and control parameters

Descriptions	Symbols	Values
Filter capacitance	C_{gf}	40 μF
Grid filter inductance	L_{gg}	5 mH
Converter filter inductance	L_{gi}	2 mH
Dc-link voltage reference	V_{gdc_ref}	400 V
Grid voltage reference (rms)	V_{s_ref}	110 V
Current reference in the *q*-axis	I_{ggq_ref}	0 A
Fundamental frequency	f_o	50 Hz
Resonance frequency	f_{cr}	665.79 Hz
Switching/sampling frequency	f_{sw}/f_s	10 kHz
Current control P gain	K_{cp}	1
Current control I gain	K_{ci}	0
Current control feedback gain	K_{id}	4
Ac voltage control P gain	$\mathbf{K_{vp}}$	**0.2/2**
Ac voltage control I gain	K_{vi}	50

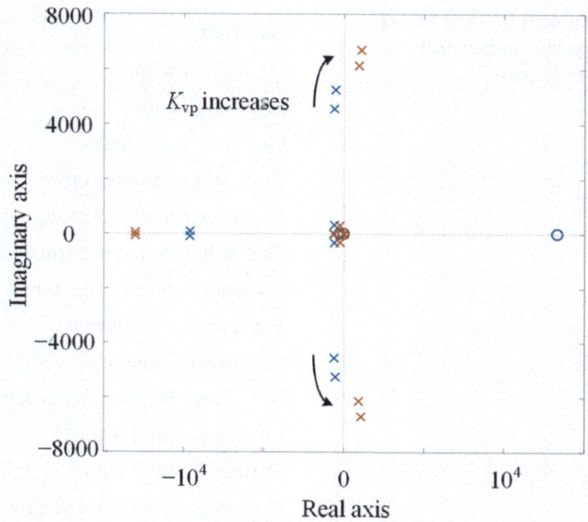

Fig. 4.16 Pole-zero maps of Type III GTCs as a function of ac voltage control P gain K_{vp}

+ repetitive regulators) as ac voltage controllers. Notably, the P gain of ac voltage controllers plays an important role on system stability. Therefore, we incorporate two different values of P gains in Table 4.4 for validation. It should be pointed out that the I gain of ac voltage controllers also affects stability, whose influence has already been analyzed in Sect. 3.2.

Figure 4.16 presents the pole-zero maps of Type III GTCs as a function of ac voltage control P gain. It is clear that the increment of K_{vp} shifts two pairs of poles from the left-half plane to the right-half plane, leading to instability. We calculate the locations of unstable poles as $1063.12 \pm 6705.64j$ and $920.04 \pm 6141.13j$. Similar to dc voltage controllers, ac voltage controllers are intertwined with ac current controllers. However, well-separated ac voltage and current controllers are not necessarily guaranteed, particularly with fast ac voltage dynamics.

4.2.1.5 Grid-Supportive Functions

On top of typical current and voltage controllers as well as grid synchronization, GTCs are increasingly expected to support power grids. Grid-supportive functions are twofold. First, GTCs can improve power quality of grids through harmonic filtering and reactive power compensation, etc., as has been discussed in Chap. 3. Second, we can use GTCs to deliver grid-supportive functions so that system-level stability is improved, which will better be grasped after the introduction of system-level stability problems in the next section. The objective of this part is to show that grid-supportive functions may cause stability problems.

Here, we use the voltage droop support delivered by Type IV GTCs as an example of demonstration. Table 4.5 documents the system and control parameters of Type IV GTCs. As observed, the dc-link capacitance C_{gdc} and dc voltage controller are

Table 4.5 Type IV GTC system and control parameters

Descriptions	Symbols	Values
Dc-link capacitance	C_{gdc}	2.82 mF
Filter capacitance	C_{gf}	40 μF
Grid filter inductance	L_{gg}	5 mH
Converter filter inductance	L_{gi}	2 mH
Dc-link voltage reference	V_{gdc_ref}	400 V
Grid voltage reference (rms)	V_{s_ref}	110 V
Current reference in the q-axis	I_{ggq_ref}	0 A
Fundamental frequency	f_o	50 Hz
Resonance frequency	f_{cr}	665.79 Hz
Switching/sampling frequency	f_{sw}/f_s	10 kHz
Current control P gain	K_{cp}	1
Current control I gain	K_{ci}	0
Current control feedback gain	K_{id}	4
Ac voltage control P gain	K_{vp}	0.2
Ac voltage control I gain	K_{vi}	50
Dc voltage control P gain	K_{dcp}	0.01
Dc voltage control I gain	K_{dci}	0.1
Reactive power filter time constant	T_q	$1/(2\pi \times 100)$
Voltage droop gain	$\boldsymbol{K_{qdroop}}$	**0/5**

involved, as Type IV GTCs are capable of forming dc grids. It should be highlighted that the voltage droop controller links the output reactive power to the ac voltage reference (in the d-axis) of GTCs, which will be detailed later. Dual to frequency droop, the voltage droop aims to automatically share reactive power among multiple GTCs [7]. Moreover, voltage droop also helps GTCs synchronize with power grids in terms of voltage amplitudes. Other grid-supportive functions that aim to improve system-level stability implemented by GTCs will be covered in the next chapter.

Figure 4.17 presents the pole-zero maps of Type IV GTCs as a function of voltage droop gain K_{qdroop}. Without voltage droop, the system is stable. As K_{qdroop} increases, the system gradually becomes unstable, characterized by a pair of unstable poles $527.09 \pm 1578.05j$.

As long as all controllers are properly designed, stability problems introduced in this subsection will disappear under normal operating conditions. However, several other factors may make us revisit these problems in nonideal conditions.

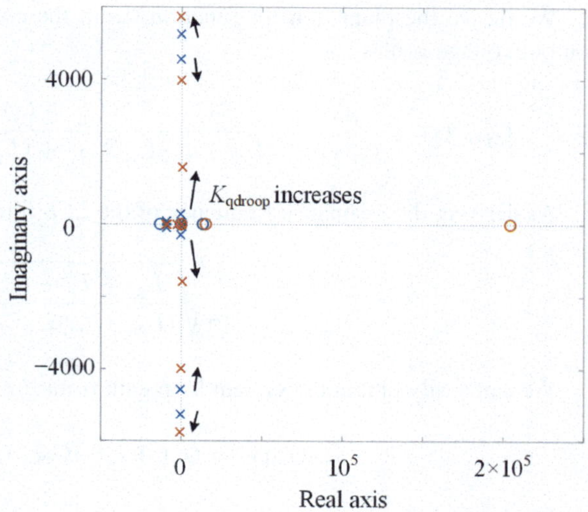

Fig. 4.17 Pole-zero maps of Type IV GTCs as a function of voltage droop gain K_{qdroop}

4.2.2 *Time Delays*

Time delays may cause stability problems. Broadly speaking, power converters experience two types of time delays—control and communication time delays. Time delays can deteriorate control, leading to stability problems mentioned in the previous subsection. However, time delays sometimes also improve stability.

4.2.2.1 Control Time Delays

As mentioned, digital control introduces time delays due to reference calculation and update. In this part, we show the effect of control time delays through an example of single-loop-controlled Type III GTCs. Such GTCs regulate ac voltages without current controllers.

We use the block-diagram model of GTCs shown in Fig. 4.18, where the time delay model remains unchanged as its original version:

$$G_{\mathrm{d}}(s) = e^{-1.5T_{\mathrm{s}}s}. \tag{4.72}$$

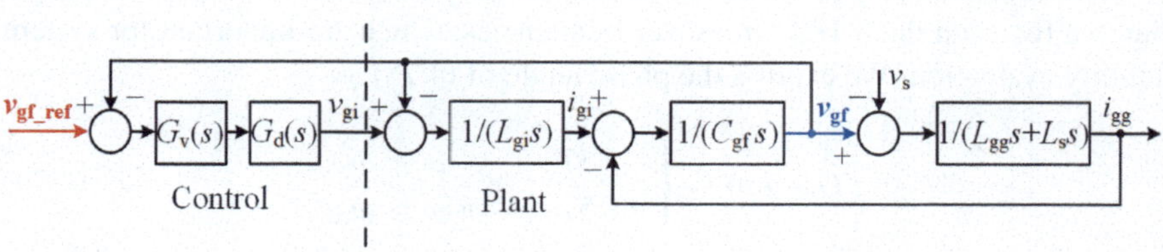

Fig. 4.18 Block diagram of single-loop-controlled Type II GTCs

We derive the plant transfer function (from the converter voltage v_{gi} to the ac output voltage v_{gf}) as

$$G_{\text{plant}}(s) = \frac{v_{\text{gf}}(s)}{v_{\text{gi}}(s)} = \frac{(L_{\text{gg}} + L_{\text{s}})}{(L_{\text{gg}} + L_{\text{s}})L_{\text{gi}}C_{\text{gf}}s^2 + (L_{\text{gg}} + L_{\text{s}} + L_{\text{gi}})}. \tag{4.73}$$

As derived, the resonance frequency of the *LCL* filter is expressed as

$$f_{\text{cr}} = \frac{1}{2\pi}\sqrt{\frac{L_{\text{gg}} + L_{\text{s}} + L_{\text{gi}}}{(L_{\text{gg}} + L_{\text{s}})L_{\text{gi}}C_{\text{gf}}}}. \tag{4.74}$$

We can easily obtain the system loop gain from Fig. 4.18 as

$$G_{\text{ol}}(s) = G_{\text{v}}(s)G_{\text{d}}(s)G_{\text{plant}}(s). \tag{4.75}$$

Further, the closed-loop voltage transfer function is expressed as

$$G_{\text{cl}}(s) = \frac{G_{\text{ol}}(s)}{1 + G_{\text{ol}}(s)}. \tag{4.76}$$

With a P controller and zero time delay, we can unfold (4.76) by using (4.73) and (4.75) as

$$G_{\text{cl}}(s) = \frac{K_{\text{vp}}(L_{\text{gg}} + L_{\text{s}})}{(L_{\text{gg}} + L_{\text{s}})L_{\text{gi}}C_{\text{gf}}s^2 + (L_{\text{gg}} + L_{\text{s}} + L_{\text{gi}}) + K_{\text{vp}}(L_{\text{gg}} + L_{\text{s}})}. \tag{4.77}$$

It is clear from (4.77) that the two closed-loop poles are on the imaginary axis, which are essentially unstable poles in practice. Therefore, Type II GTCs with P controllers as their single-loop voltage regulators will be unstable without any time delay.

Next, we substitute the time delay model in (4.72) and the plant model in (4.73) into (4.75). It yields

$$G_{\text{ol}}(s) = \frac{K_{\text{vp}}(L_{\text{gg}} + L_{\text{s}})e^{-1.5T_{\text{s}}s}}{(L_{\text{gg}} + L_{\text{s}})L_{\text{gi}}C_{\text{gf}}s^2 + (L_{\text{gg}} + L_{\text{s}} + L_{\text{gi}})}. \tag{4.78}$$

To conduct stability analysis, we should identify the phase of $G_{\text{op}}(s)$. In particular, we focus on the $-180°$ crossover frequencies, which are important for system stability evaluation. We express the phase angle of (4.78) as

$$\angle G_{\text{ol}}(j\omega) \approx \begin{cases} -1.5T_{\text{s}} & \omega < \omega_{\text{cr}} \\ -1.5T_{\text{s}}\omega - \pi & \omega > \omega_{\text{cr}} \end{cases}, \tag{4.79}$$

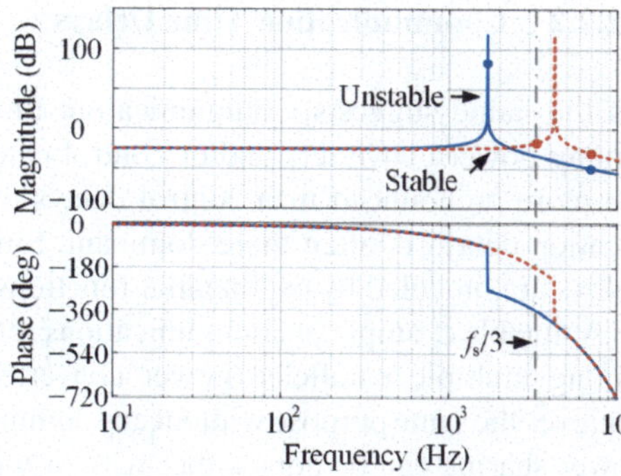

Fig. 4.19 Block diagram of single-loop-controlled Type II GTCs

where $\omega_{cr} = 2\pi f_{cr}$ denotes the resonance angular frequency. We see from (4.79) that a $-\pi$ (or $-180°$) phase shift occurs at ω_{cr}.

The Nyquist stability criterion tells that a stable system should feature zero $-180°$ crossing if its system loop gain contains zero right-half-plane pole [8]. Referring to (4.78), we confirm that $G_{ol}(s)$ has zero right-half-plane pole. For stable systems, we should guarantee that the phase angle at ω_{cr} is in the range of $(-2k\pi)$ to $(-2k\pi + \pi)$, where k is an integer. That is,

$$(-2k+1)\pi > -1.5T_s\omega_{cr} > -2k\pi \longrightarrow \frac{2f_s}{3} > f_{cr} > \frac{f_s}{3}, \tag{4.80}$$

where $k = 1$ is chosen to yield the main resonance frequency range.

Figure 4.19 illustrates the Bode diagrams of $G_{ol}(s)$ with two different f_{cr}, in which K_{vp} is designed as $0.001/(L_{gg} + L_s)$. We notice resonant peaks in both cases. In the case of $f_{cr} < f_s/3$ (solid curve), we find a negative $-180°$ crossing at f_{cr}, indicating an unstable system. In the other case of $2f_s/3 > f_{cr} > f_s/3$ (dotted curve), there is no zero crossing, showing that the system is stable. In this case, time delays extend stability regions, making $(2f_s/3 > f_{cr} > f_s/3)$ a stable region.

From the above discussions, we conclude that control time delays may positively contribute to system stability. However, it should be emphasized that time delays in fact deteriorate system stability in most scenarios. For example, control time delays shrink the stability region of single-loop-controlled converter-current-feedback GTCs, as detailed by comparison of [9, 10]. Notably, the inverse model of time delays cannot be directly implemented in controllers to compensate time delays. Even an approximation of the inverse time-delay model, such as $(1 + 1.5T_s s)$, is difficult to implement in practice due to the amplification of high-frequency noise. We will discuss relevant solutions in the next chapter.

4.2.2.2 Communication Time Delays

As the name suggests, communication time delays come from communications among power converters and/or control centers. Generally, communication delays are more pronounced than control delays. Correspondingly, the effect of communication delays is often more dominant. Similar to control delays, communication delays are modeled by exponential functions (or their Padé approximants).

A notable example of communications refers to active and reactive power sharing among multiple paralleled power converters. As mentioned, droop control also achieves the same purpose without any communication. Generally, active and reactive power sharing can be done either with or without communications. In comparison, power sharing through communication is mature and flexible at the expense of extra costs and time delays. In practice, the majority of power systems features communications (at least of low-speed). When local control (e.g., current/voltage control) is achieved without communication, while communications are applied to high-level control (e.g., power reference control), we obtain a standard hierarchical control architecture in practical power systems.

Taking communication time delays into consideration, we can conduct stability analysis similarly to the previous part, and hence not repeated here. When GTCs deliver grid-supportive services, the analysis of communication time delays on system stability has been detailed in [11]. We briefly show the relevant experimental results here.

Figure 4.20 presents the experimental waveforms of GTCs with/without communication time delays, where τ denotes the length of time delays [11]. Without any communication time delay, GTCs operate normally. Along with a time delay of 5 s is a system instability, which triggers relays for protection.

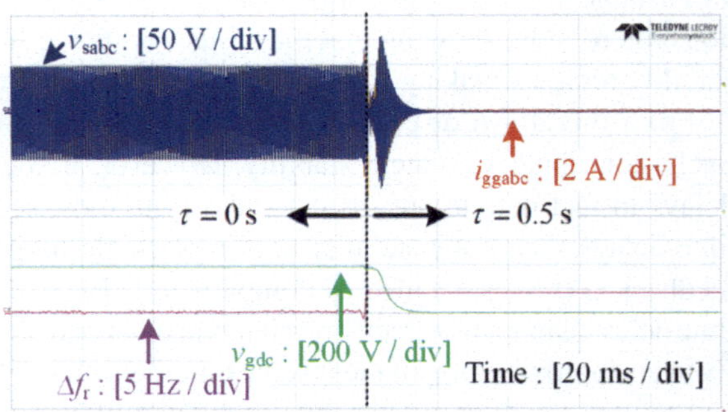

Fig. 4.20 Experimental waveforms of GTCs with/without communication time delays [11]

4.2.3 *Weak Grids*

Weak-grid-induced stability problems have challenged the operation of GTCs for more than 15 years. Preliminary research regarding the effect of weak grids on stability can be found in [12].

As mentioned in Sect. 1.3, power grids contain power lines. According to the nature of line impedances, we classify power grids into stiff (or strong) and weak grids. Stiff grids have small line impedances with negligible effects. As a result, stiff grids are simply modeled as ideal ac voltage sources. In contrast, we have to consider line impedances in weak grids, which often feature large values and may vary greatly according to the changes of grid configurations and operating conditions. We have modeled transmission lines as distributed parameter circuits in Sect. 1.3, which can further be simplified into series *RL* circuits. As the voltage level increases, the inductive component becomes dominant. As such, weak grids are often modeled by serial connections of inductors and ac voltage sources. To evaluate the strength of grids, the metric SCR has been defined in (1.99). Notably, GTCs can only access the PCC instead of real grid voltages in weak grids.

The interaction between weak grids and converter control introduces stability problems to GTCs. In this book, we propose to classify the mechanisms of weak-grid-induced stability problems into two aspects—parameter changes and local control loops. We will explore them separately.

4.2.3.1 Parameter Changes

In the plant models of GTCs [see (4.5)–(4.8)], we find the line inductance L_s. Therefore, the change of L_s can alter plant parameters, thereby affecting the stability of GTCs. A vivid example of parameter changes refers to the shift of filter resonance frequencies. If resonance frequencies are shifted out of allowable ranges, GTCs will be unstable [13]. For *L*-filtered GTCs, the change of L_s also modifies system plants.

As a straightforward example, an *L*-filtered Type I GTC is selected. Its system loop gain can mathematically be described by (4.75), where the respective transfer functions are changed as

$$G_v(s) = K_{cp} + \frac{K_{ci}}{s}, \tag{4.81}$$

$$G_d(s) = \frac{1 - 0.75T_s s}{1 + 0.75T_s s} \approx e^{-1.5sT_s}, \text{ and} \tag{4.82}$$

$$G_{plant}(s) = \frac{1}{s(L_{gg} + L_s)} = \frac{1}{sL_t}. \tag{4.83}$$

Correspondingly, the closed-loop transfer function $G_{cl}(s)$ is also expressed as (4.76).

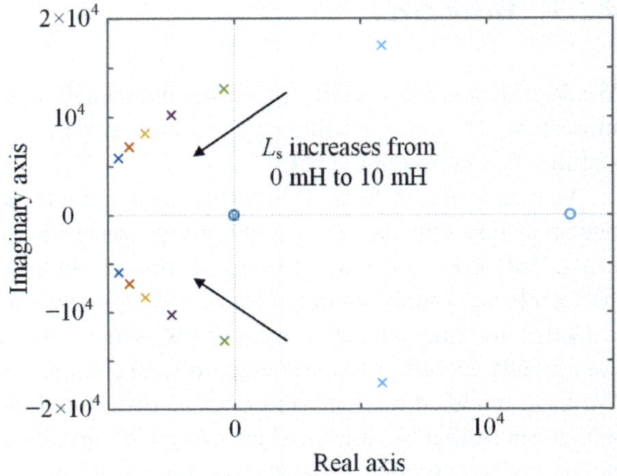

Fig. 4.21 Pole–zero map of $G_{cl}(s)$ as a function of the line inductance L_s

Figure 4.21 illustrates the pole–zero map of $G_{cl}(s)$ as a function of L_s, where L_s ranges from 0 to 10 mH. Clearly, the change of L_s can affect system stability. In this case, we notice that the increment of L_s contributes positively to system stability. In this sense, the influence of weak grids on stability can be a double-edged sword, which is very similar to that of time delays. As L_s increases, the gain of $G_{plant}(s)$ decreases, thus shifting the unstable poles leftwards to the stable regions in the left-hand plane.

4.2.3.2 Local Control Loops

Weak grids may introduce undesirable local control loops that cause instability. For example, weak grids, in combination with PLLs, can yield local loops that pose challenges to grid synchronization [3]. Besides, weak grids can destabilize GTCs that deliver grid-supportive services through additional control loops [5].

Once again, we employ Type I GTCs for demonstration. In this case, we incorporate the model of PLLs into the control block diagram of GTCs. Figure 4.22 demonstrates the control block diagram of Type I GTCs in weak grids, where the capital letters denote the steady-state values of relevant variables.

In Fig. 4.22, $G_{pll}(s)$ models the transfer function of PLLs in Fig. 4.3 (without filters for simplification) as

$$G_{pll}(s) = \frac{\Delta\theta_{pll}(s)}{\Delta v_{gq}(s)} = \frac{K_{pllp}s + K_{plli}}{s^2 + V_{gd}K_{pllp}s + V_{gd}K_{plli}}. \tag{4.84}$$

We observe from Fig. 4.22 that there appears an additional local control loop introduced by weak grids and PLLs (through L_s/L_t, $G_{pll}(s)$, and I_{ggd} or V_{gd}). If GTCs also compensate reactive currents, the local control loop will link to the d-axis. We assume that $I_{ggq} = 0$ A and focus only on the q-axis to investigate the effect of this local control loop.

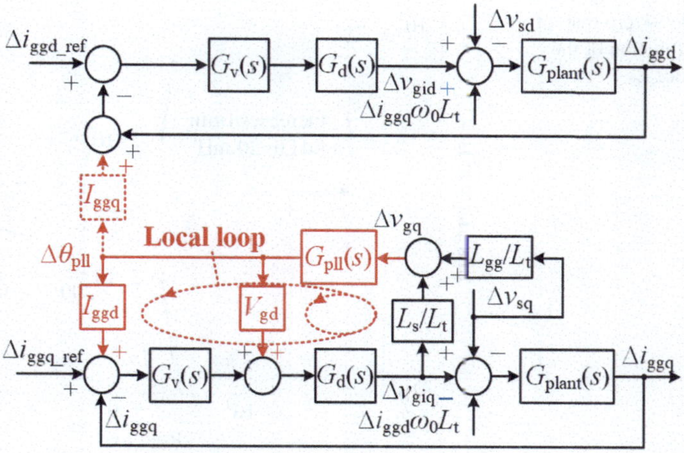

Fig. 4.22 Control block diagram of Type I GTCs in weak grids

Ignoring the coupling effect, we derive the q-axis closed-loop current control transfer function as

$$G_{\mathrm{clq}}(s) = \frac{\Delta i_{\mathrm{ggq}}(s)}{\Delta i_{\mathrm{ggq\_ref}}(s)} = \frac{G_{\mathrm{v}}(s)G_{\mathrm{dm}}(s)G_{\mathrm{plant}}(s)}{1 + G_{\mathrm{v}}(s)G_{\mathrm{dm}}(s)G_{\mathrm{plant}}(s)}, \tag{4.85}$$

where

$$G_{\mathrm{dm}}(s) = \frac{G_{\mathrm{d}}(s)}{1 - \left[I_{\mathrm{ggd}}G_{\mathrm{v}}(s) + V_{\mathrm{gd}}\right]G_{\mathrm{pll}}(s)G_{\mathrm{d}}(s)L_s/L_t}. \tag{4.86}$$

Note that $G_{\mathrm{dm}}(s)$ models the local control loop.

Figure 4.23 presents the pole–zero map of $G_{\mathrm{cq}}(s)$ as a function of L_s. For clarity, we reduce the current control proportional gain K_{cp} so that the system becomes stable under the condition of $L_s = 0$ mH. Under this condition, the increment of L_s destabilize GTCs, as one pair of unstable poles emerges. Notably, the change of L_s affects both $G_{\mathrm{plant}}(s)$ and $G_{\mathrm{dm}}(s)$ in (4.85), and their effects on system stability are opposite. Although the change of $G_{\mathrm{plant}}(s)$ contributes to stability (see Fig. 4.21), the local control loop $G_{\mathrm{dm}}(s)$ plays a negative yet dominant role.

In addition to PLLs, weak grids may also introduce local control loops (and hence instability) to GTCs with grid-supportive services. We will present the relevant phenomenon and solutions in the next chapter.

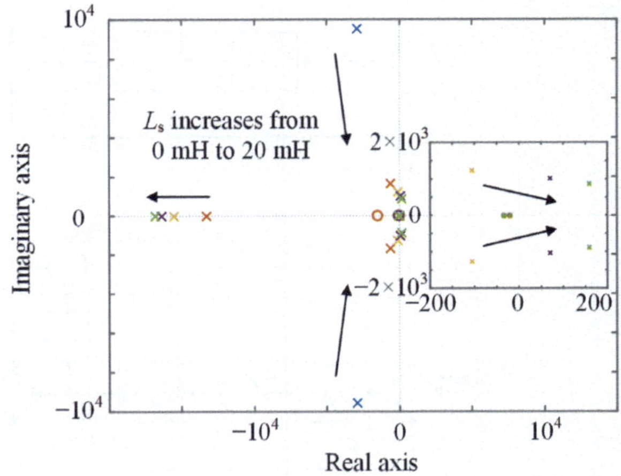

Fig. 4.23 Pole–zero map of $G_{clq}(s)$ as a function of the line inductance L_s

4.2.4 *Power Quality Problems*

Without prompt treatments, power quality problems may further develop into stability problems. As detailed in Chap. 2, GTCs face typical voltage quality problems including voltage amplitude deviations, voltage harmonics, and voltage imbalances. We will separately discuss their effects on stability.

4.2.4.1 Voltage Amplitude Deviations

Voltage amplitude deviations can destabilize GTCs in two ways. First, voltage drops lead to large currents under power control (or voltage control) modes, which may exceed the current rating of GTCs. Second, voltage deviations change the operating points of GTCs, which can be unstable points.

To illustrate, Fig. 4.24 shows the simulation results of unstable Type I GTCs under voltage sags, where currents become distorted during voltage sags. It should be

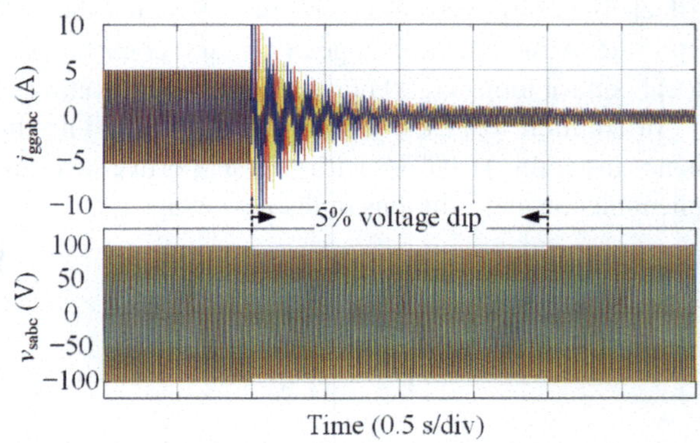

Fig. 4.24 Simulation results of Type I GTCs under voltage sags

remembered that voltage amplitude deviations often greatly change system operating points. As a result, stability analysis at a single operating point will be insufficient. In this case, we can employ large-signal stability analysis techniques, as presented in [14]. An alternative option is robust stability analysis, in which we consider parameter variation ranges.

4.2.4.2 Voltage Harmonics

Voltage harmonics will be amplified by GTCs when a positive feedback appears at harmonic frequencies. Conventionally, the relevant harmonic distortions are regarded as power quality problems. However, harmonics may complicate system models and cause stability problems through frame rotations.

Referring to Table 2.3 (in Sect. 2.1), we notice that different-order harmonics in the natural frame (e.g., 2nd and 4th) can be mapped into the same-order harmonic (e.g., 3rd). Therefore, the relevant transfer function (at the frequency of third harmonic) may be distorted by such multiple inputs. Alternatively, the same-order harmonic in the *dq*-frame can be mapped into different-order harmonics in the natural frame (see Table 2.4). The coupling of multiple harmonics complicates system models and stability analysis, as detailed in [15, 16].

4.2.4.3 Voltage Imbalances

Similar to voltage harmonics, voltage imbalances complicate GTC models through frame rotations. To handle voltage harmonics and imbalances, a popular solution is to build system models for individual sequences and harmonic orders. On top of that, the coupling effects among such models should be considered. This method is known as harmonic state-space modeling [16]. Through this approach, the number of state variables first triples as single harmonics and then multiplies by the number of harmonics. Therefore, the resultant system model can be very accurate at the expense of great complexity.

Up to now, we have finished the introduction of converter-level stability problems. To summarize, Table 4.6 classifies stability problems according to their sources and time scales, where time scales are divided according to the fundamental frequency f_o. Specifically, the three ranges ($\ll f_o$), ($\approx f_o$), and ($\gg f_o$) stand for sub-fundamental, near-fundamental, and super-fundamental frequencies, respectively. Notably, active and reactive power support are grid-supportive services, which will further be discussed in the next chapter. Clearly, improper controller designs, time delays, weak grids, and power quality problems may give rise to almost all converter-level stability problems. Another thing to note is that stability problems can be dependent on GTC types. For example, ac grid-following GTCs (Types I and II GTCs) employ grid synchronization units (such as PLLs) but without any ac voltage controller.

Table 4.6 Converter-level Stability problems

Stability problems	Sources	Time scales
Active power support	Improper controller designs, time delays, weak grids, power quality problems	$\ll f_o$, $\approx f_o$, or $\gg f_o$
Reactive power support	Improper controller designs, time delays, weak grids, power quality problems	$\ll f_o$, $\approx f_o$, or $\gg f_o$
Dc voltage control (dc grid forming)	Improper controller designs, time delays, weak grids, power quality problems	$\ll f_o$, $\approx f_o$
Ac voltage control (ac grid forming)	Improper controller designs, time delays, weak grids, power quality problems	$\approx f_o$, or $\gg f_o$
Grid synchronization (ac grid following)	Improper controller designs, weak grids, power quality problems	$\ll f_o$, $\approx f_o$
Ac current control	Improper controller designs, time delays, weak grids, power quality problems	$\gg f_o$

4.3 System-Level Stability Problems

System-level stability problems are mainly caused by active and/or reactive power imbalances. Generally, active power balance is closely related to the system frequency, while the voltage amplitude reflects a degree of balance for reactive power. In conventional power systems, power balance, particularly for active power balance, is predominantly maintained by SGs. In more-electronics power systems, GTCs and SGs collectively regulate power and guarantee system-level stability.

4.3.1 Active Power Balance Problems

We have detailed the modeling of active power generation in Sect. 1.3. Returning to Fig. 1.43 (or Fig. 1.31), we note that the frequency f_{r_pu} (or the rotor speed of SGs) is determined by the generated active power reference p_{ref_pu} (or the mechanical torque of SGs) and the demanded power p_{load_pu} (or the electrical torque of SGs). Therefore, active power influences the mains frequency, and vice versa. As the integral of the frequency, the phase angle of SGs or GTCs also has a tight bearing on active power balance. Active power imbalances can cause frequency and/or angle stability problems.

4.3.1.1 Frequency Stability Problems

Frequency control is a priority for the majority of power system operators. This is because low-frequency operations can mechanically overstress the turbine blades or overheat boilers of SGs due to reduced cooling from pumps and fans [7]. Therefore, under-frequency relays are used in power systems to protect SGs. If the frequency drops too much, SG tripping, cascading failures, and/or even large-scale blackouts may occur. To avoid blackouts, many power system operators prescribe standards on frequency control, as listed in Table 4.7 [17]. Notably, some countries (e.g., Singapore) fix the thresholds for low-frequency load shedding.

Under normal operations, the grid frequency should be maintained within the normal bands listed in Table 4.7. In the face of frequency contingencies (e.g., a sudden loss of SGs or increment of loads), frequency variations should not exceed the tolerant bands in Table 4.7.

Figure 4.25 sketches a frequency response curve during a generator tripping event, where the primary frequency control (in the time scale of seconds) arrests the fast frequency decline, and then the secondary frequency control restores the frequency to its nominal value (in several or tens of minutes) [17].

There are two important frequency indices in Fig. 4.25. The first one is the frequency nadir, which refers to the lowest frequency point. Notably, the frequency nadir must be controlled within its tolerant bands (see Table 4.7). The second one is the rate of change of frequency df_g/dt (denoted as RoCoF). Excessive RoCoF levels cause fast frequency changes, which may lead to pole slipping of SGs and catastrophic

Table 4.7 Selected frequency control standards [17]

Countries	Rated frequency (Hz)	Normal bands (Hz)	Tolerant bands (Hz)
Australia	50	±0.15	±1
China	50	±0.2	±1
Europe	50	±0.2	±0.8
Singapore	50	±0.2	−0.3
UK	50	±0.5	±1

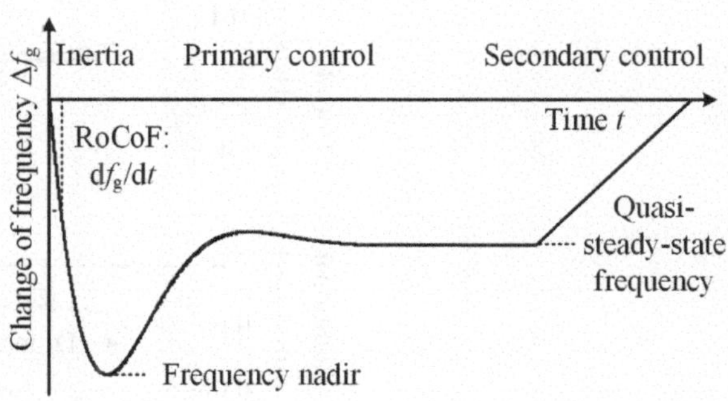

Fig. 4.25 Frequency response following a SG tripping event [17]

results. For protection, SGs often employ RoCoF relays so that they will be tripped under excessive RoCoF levels (e.g., 1.5 to 2 Hz/s over a 500 ms rolling window) [18]. In addition, many anti-islanding protective relays are designed as RoCoF relays [17]. As a result, excessive RoCoF levels can lead to SG tripping, system protection, and/or system blackouts. Some power system operators also prescribe standards on RoCoF withstand capabilities, such as 1 Hz/s for 0.5 s [18]. In general, frequency control targets at the regulation of frequency nadir and RoCoF. Next, we proceed to the two emerging challenges of frequency control.

Damping Shortage Challenge

Recapping the model of frequency regulation in Fig. 1.43, we have defined the lumped damping factor D. In fact, D plays an essential role on the stability of frequency control.

To illustrate, Fig. 4.26 shows four primary frequency response curves under contingencies, where the curves are plotted based on the block diagram in Fig. 1.43. As compared with the baseline case, the case of increased damping D obviously improves frequency control in terms of the frequency nadir. In addition, the quasi-steady-state frequency (after primary response) is also improved. It should be emphasized that D has no direct effect on the RoCoF, particularly at the instant of contingencies.

In addition to baseline and damping, we observe two other cases, which illustrate the effects of droop and inertia, respectively. We will discuss on inertia later and focus on droop now. As compared to damping, frequency droop responds more slowly, and hence features a lower (or poorer) frequency nadir. However, the two cases are of identical quasi-steady-state frequencies. Reviewing Fig. 1.43 once more (where the damping factor D can be treated as a feedback gain that links to the load node), we find that the damping factor D and droop gain K_{droop} both represent the proportional

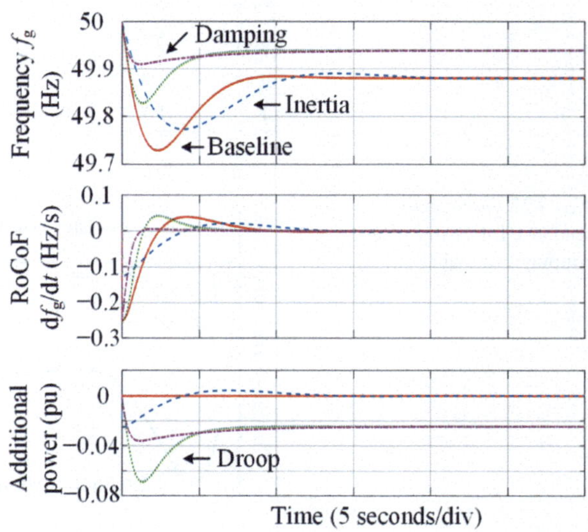

Fig. 4.26 Primary frequency response curves under contingencies

changes of active power as a function of the frequency change. The difference is that damping directly acts on the load node, while droop changes the active power reference. Assuming that speed governors and turbines operate very fast (with their dynamics ignored, i.e., $G_{\text{Governor}}(s) = G_{\text{Turbine}}(s) = 1$), we can conclude that droop and damping are equivalent. In this sense, frequency droop can be regarded as a type of slow damping.

Since damping improves frequency nadirs, the frequency may alternatively go beyond acceptable ranges (listed in Table 4.7) without sufficient damping. As mentioned, damping is mainly contributed by SGs and loads (particularly for induction motor loads) [7]. In more-electronics power systems, power converters gradually replace SGs. Moreover, induction motor loads are increasingly connected to the grid through GTCs for better speed regulation. Both cases lead to a damping shortage challenge. The damping shortage deteriorates frequency nadirs, which in combination with stochastic renewable outputs, pose a serious challenge to frequency control in more-electronics power systems.

Inertia Shortage Challenge

As observed in Fig. 1.43, the inertia coefficient H_{SG} has been defined as the combined (per unit) kinetic energy of SGs in Chap. 1. Therefore, inertia mainly come from SGs. Importantly, inertia plays a significant role on the stability of frequency control.

We refer to Fig. 4.26 for the effect of inertia. As compared with the baseline case, the case of increased inertia features significant improvements in both frequency nadir and RoCoF. However, inertia does not affect the quasi-steady-state frequency. To understand this, we can recap Fig. 1.43 and the swing equation in Chap. 1. Notably, the power difference between generation and demand equals $H_{\text{SG}} \mathrm{d}f_{\text{g}}/\mathrm{d}t$. With a fixed power difference, a larger inertia coefficient translates into a smaller RoCoF. As such, inertia affects the RoCoF directly. In quasi-steady state, the frequency becomes fixed (i.e., $\mathrm{d}f_{\text{g}}/\mathrm{d}t = 0$), thereby indicating that inertia has no effect. In other words, inertial power has fast dynamics and functions in the beginning of frequency events, as shown in Figs. 4.25 and 4.26. The quantitative analysis of inertia effects (on RoCoF, nadir, etc.) based on Fig. 1.43 is detailed in [19]. In addition, common values of H are 5 s for gas-fired generators, 3.5 s for coal-fired generators, 4 s for nuclear generators, and 3 s for hydraulic generators [17].

In more-electronics power systems, SGs are replaced by power converters, leading to an inertia shortage challenge. This challenge is so significant that power system researchers put the operation of low-inertia power systems as one of the four major topics in their annual meeting. Also, this challenge is highly valued by the industry. As an example, it is concluded by the Australian Energy Market Operator in its final report that the lack of inertia is a reason for the 2016 Blackout [20]. We will detail the methods of inertia enhancement in the next chapter.

4.3.1.2 Angle Stability Problems

In conventional power systems, stability mainly refers to the synchronous (or in-step) operation of SGs [7]. To ensure stability, phase angles and power exchanges among SGs should be maintained stable. Such stability lies in two aspects. First, we should avoid continuous phase angle oscillations, such as sub-synchronous oscillations [7]. Second, we must guarantee that active power delivered by individual SGs and GTCs are within their boundaries.

Phase Angle Oscillations

In more-electronics power systems, phase angle oscillations occur without proper damping and/or control. To investigate phase angle oscillations, we first model GTCs and SGs (respectively given in Sects. 4.1 and 1.3) as well as based on their interconnection relationships. After that, we analyze the state-space model of the lumped systems consisting of multiple SGs and GTCs, based on which we further tune controllers. In general, the analysis methodology remains unchanged, and hence will not be repeated here.

Excessive Phase Angles

Before investigating stability problems due to excessive phase angles, we first derive the relationship between active power and phase angle through a simplified voltage-source model of two interconnected SGs and/or GTCs.

Figure 4.27 illustrates the schematic of two interconnected voltage sources, where the phase angle of the first voltage source is selected as a reference of zero degree. V_1 and V_2 represent the voltage amplitudes of two voltage sources, respectively. In particular, we focus on the phase angle δ, which denotes the angle difference between the two voltage sources. In addition, I_{12} and ϕ designate the amplitude and phase angle of the current phasor, respectively. The line impedance is expressed as

$$Z_{12}\angle\theta = R_{12} + jX_{12}, \tag{4.87}$$

where R_{12} and X_{12} stand for the line resistance and reactance, respectively.

For clarity, Fig. 4.28 presents the phasor diagram corresponding to Fig. 4.27, where the relationships among various vectors are visualized. We reorganize the current phasor as

$$I_{12}\angle - \phi = \frac{V_1\angle 0^\circ - V_2\angle - \delta}{Z_{12}\angle\theta}. \tag{4.88}$$

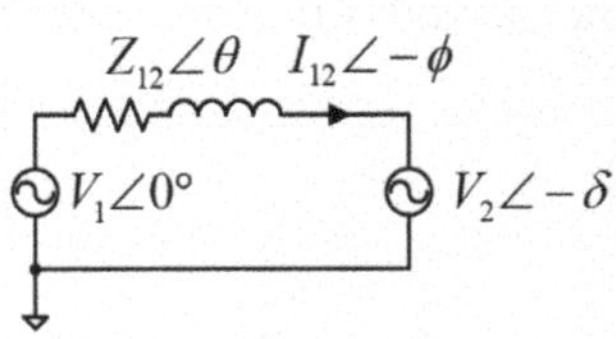

Fig. 4.27 Schematic of two interconnected voltage sources

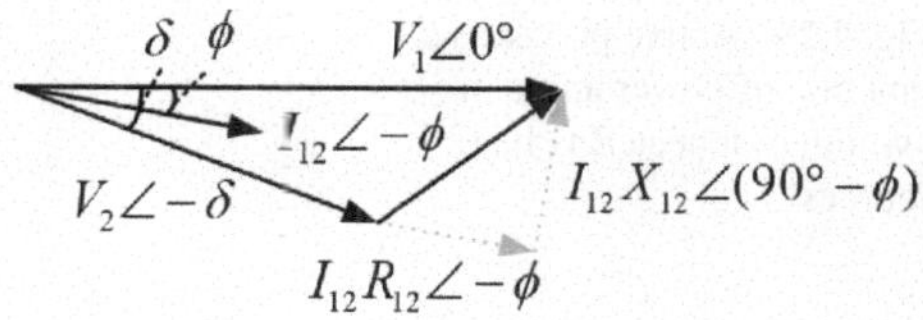

Fig. 4.28 Phasor diagram of two interconnected voltage sources

Substitution of (4.87) into (4.88), the latter yields

$$I_{12}\angle-\phi=\frac{V_1-V_2\cos\delta+jV_2\sin\delta}{R_{12}+jX_{12}}$$
$$=\frac{R_{12}(V_1-V_2\cos\delta)+X_{12}V_2\sin\delta}{R_{12}^2+X_{12}^2}+j\frac{R_{12}V_2\sin\delta-X_{12}(V_1-V_2\cos\delta)}{R_{12}^2+X_{12}^2}. \quad (4.89)$$

The exchanged complex power (from the source 1 to the source 2) takes the form of

$$\mathbf{S_{12}}=\frac{1}{2}V_1I_{12}\angle\phi. \quad (4.90)$$

Substituting of (4.89) into (4.90), we expand the complex power as

$$\mathbf{S_{12}}=\frac{R_{12}V_1(V_1-V_2\cos\delta)+X_{12}V_1V_2\sin\delta}{2(R_{12}^2+X_{12}^2)}+j\frac{X_{12}V_1(V_1-V_2\cos\delta)-R_{12}V_1V_2\sin\delta}{2(R_{12}^2+X_{12}^2)}. \quad (4.91)$$

Furthermore, we obtain the active and reactive power from (4.91) as

$$P_{12}=\frac{R_{12}V_1(V_1-V_2\cos\delta)+X_{12}V_1V_2\sin\delta}{2(R_{12}^2+X_{12}^2)} \text{ and} \quad (4.92)$$

$$Q_{12}=\frac{X_{12}V_1(V_1-V_2\cos\delta)-R_{12}V_1V_2\sin\delta}{2(R_{12}^2+X_{12}^2)}. \quad (4.93)$$

The above equations are very helpful for demonstration of angle stability limits.

In this part, we concentrate on active power. For purely inductive lines (i.e., $R_{12} = 0\ \Omega$), we simplify the active power expression as

$$P_{12}=\frac{V_1V_2\sin\delta}{2X_{12}}=K_{12}\sin\delta, \quad (4.94)$$

where K_{12} stands for the coefficient of active power transfer. Obviously, the active power P_{12} is proportional to the sinusoidal of the phase angle δ (also known as the power angle).

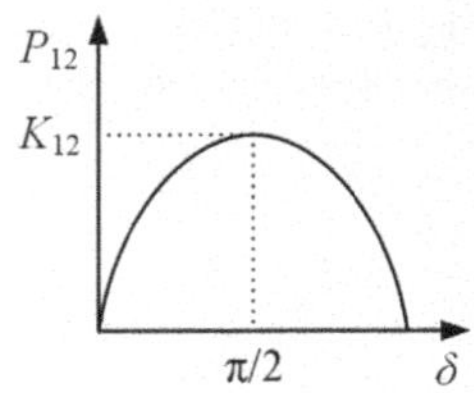

Fig. 4.29 Active power as a function of power angle of two interconnected voltage sources

Figure 4.29 visualizes the power-angle relationship of two voltage sources. Clearly, the active power reaches its peak when the angle δ equals $\pi/2$. Meanwhile, the power-angle relationship changes from positive to negative as δ increases and exceeds $\pi/2$. Notably, for SGs and ac grid-forming GTCs, we assume a positive relationship between the active power and phase angle. If the active power output is lower than its reference, the SG or GTC will increase δ to control active power. When δ exceeds 90°, the increment of δ will lead to the reduction of active power, thus introducing a positive feedback and instability. To avoid stability problems caused by excessive phase angles, we should maintain phase angle differences among SGs and GTCs within $\pi/2$ (or 90°). Correspondingly, the maximum active power K_{12} is an upper limit of active power transfer, which depends on the line impedance and voltage amplitudes.

In the case of n voltage sources, their active power-angle relationships are described by the following nonlinear equations:

$$P_{\mathrm{ij}} = \frac{V_{\mathrm{i}} V_{\mathrm{j}} \sin \delta_{\mathrm{ij}}}{2X_{\mathrm{ij}}} = K_{\mathrm{ij}} \sin \delta_{\mathrm{ij}} (i \neq j), \; i, j \in 1, 2, \ldots, n. \tag{4.95}$$

To ensure stability, we should maintain all the angles δ_{ij} within 90°.

Stability analysis of multiple controlled voltage sources (or coupled oscillators) is quite complicated. Nevertheless, when the model of each voltage source is simplified as a swing equation (i.e., Assumption I), the relevant analysis becomes tractable. In this case, we model system dynamics as

$$2H_{\mathrm{i}} \frac{\mathrm{d}^2 \delta_{\mathrm{i}}}{\mathrm{d}t^2} = T_{\mathrm{i}} - D_{\mathrm{i}} \frac{\mathrm{d}\delta_{\mathrm{i}}}{\mathrm{d}t} - \sum_{j=1}^{n} K_{\mathrm{ij}} \sin(\delta_{\mathrm{i}} - \delta_{\mathrm{j}}), \; i \in 1, 2, \ldots, n. \tag{4.96}$$

In fact, (4.96) is a second-order Kuramoto model with inertia and damping [21]. Under the assumption of negligible inertia (i.e., Assumption II), we simplify (4.96) as

$$D_{\mathrm{i}} \frac{\mathrm{d}\delta_{\mathrm{i}}}{\mathrm{d}t} = T_{\mathrm{i}} - \sum_{j=1}^{n} K_{\mathrm{ij}} \sin(\delta_{\mathrm{i}} - \delta_{\mathrm{j}}), \; i \in 1, 2, \ldots, n. \tag{4.97}$$

If all line inductances are identical (i.e., Assumption III), we further simplify the model into

$$\frac{d\delta_i}{dt} = T_i - K\sum_{j=1}^{n}\sin(\delta_i - \delta_j),\ \ i \in 1, 2, \ldots, n. \tag{4.98}$$

The above model is a classic and standard Kuramoto model, where $K > 0$ is defined as the coupling strength [21]. It should be highlighted that the stability condition of (4.96)–(4.98), in which all phase angle differences stary within $\pi/2$, has been derived through the algebraic graph theory in [22].

Importantly, we have assumed purely inductive lines in the above discussions. However, this is not necessarily satisfied in low-voltage networks, where resistive components are prominent. In addition, since stability problems related to excessive phase angles are normally encountered during system transients, they are also classified into transient stability problems in conventional power systems [7].

4.3.2 Reactive Power Balance Problems

As mentioned in Chap. 2, reactive power compensation is in the field of power quality. Also, reactive power mismatches cause system-level stability problems. In this sense, reactive power conditioning equipment (e.g., STATCOMs, APFs, and UPQCs) can contribute to system-level stability improvement.

To illustrate, we simplify (4.93) by considering purely inductive lines as

$$Q_{12} = \frac{V_1(V_1 - V_2\cos\delta)}{2X_{12}}. \tag{4.99}$$

With a minor phase angle δ, the above equation is further simplified into

$$Q_{12} \approx \frac{V_1(V_1 - V_2)}{2X_{12}} = \frac{V_1\Delta V_{12}}{2X_{12}}, \tag{4.100}$$

where the reactive power Q_{12} is positively related to the voltage amplitude difference ΔV_{12}. If this positive relationship changes into a negative one, stability problems occur. As an example, we assume constant V_1 ($V_1 > V_2$) and explore the case of $\delta > \pi/2$ (i.e., $\cos\delta < 0$). In this case, the increment of ΔV_{12} (i.e., the decrease of V_2) reduces Q_{12}, leading to instability.

4.3.2.1 Voltage Stability Problems

Voltage stability problems refer to excessive voltage amplitude deviations, which are mainly caused by reactive power mismatches. Although voltage stability problems are local phenomenon, they often introduce a system-wide impact [7].

To illustrate voltage stability problems, we consider a simplified source-load model in Fig. 4.30, where the current magnitude can be derived as

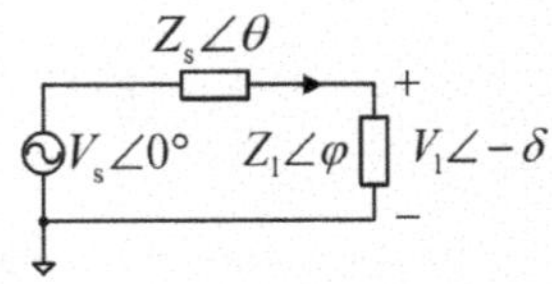

Fig. 4.30 Simplified source-load model

$$I_l = \frac{V_s}{\sqrt{(Z_s\cos\theta + Z_l\cos\varphi)^2 + (Z_s\sin\theta + Z_l\sin\varphi)^2}} = \frac{V_s}{Z_s\sqrt{F_{sl}}}, \tag{4.101}$$

where [7]

$$F_{sl} = 1 + \frac{Z_l^2}{Z_s^2} + 2\frac{Z_l}{Z_s}\cos(\theta - \varphi). \tag{4.102}$$

The magnitude of the load voltage takes the form of

$$V_l = I_l Z_l = \frac{Z_l V_s}{Z_s\sqrt{F_{sl}}}. \tag{4.103}$$

Moreover, the active power delivered to the load is

$$P_l = I_l^2 \mathrm{Re}(Z_l\angle\varphi) = \frac{V_s^2 Z_l}{Z_s^2 F_{sl}}\cos\varphi. \tag{4.104}$$

The maximum active power delivery is achieved when the load matches the line in terms of impedances, i.e.,

$$Z_l\angle\varphi = Z_s\angle\theta. \tag{4.105}$$

In this case, the maximum power equals

$$P_{l\_max} = \frac{V_s^2}{4Z_s}\cos\varphi. \tag{4.106}$$

We note that there is an upper limit of active power. Generally, we reduce the load to increase its active power. However, if the relevant active power reaches its upper limit, voltage stability problems will occur in terms of voltage collapses [7]. Note that the increment of reactive power transfer deteriorates voltage limits. From the above discussions, we learn that voltage stability problems often go hand in hand with active power transfer limits and excessive angle problems.

Until now, we have discussed system-level stability problems. System-level stability problems can be in the range of seconds or tens of seconds. It is worthwhile to note that long-time active and reactive power balances (e.g., in the minute scale) are in the field of optimization, although some textbooks classify them into long-term stability. Such problems can be investigated through simulation software.

In this chapter, we have discussed stability problems in more-electronics power systems. In general, stability problems are classified into converter-level and system-level stability problems. The sources of converter-level stability problems include improper controller designs, time-delays, weak grids, and power quality problems. Additionally, we divide system-level stability problems into active and reactive power imbalances. Through stability analyses, we disclose the mechanisms behind instability.

References

1. Rocabert J, Luna A, Blaabjerg F, Rodriguez P (2012) Control of power converters in AC microgrids. IEEE Trans Power Electron 27(11):4734–4749
2. Chung S (2000) A phase tracking system for three phase utility interface inverters. IEEE Trans Power Electron 15(3):431–438
3. Fang J, Li X, Li H, Tang Y (2018) Stability improvement for three-phase grid-connected converters through impedance reshaping in quadrature-axis. IEEE Trans Power Electron 33(10):8365–8375
4. Wen B, Boroyevich D, Burgos R, Mattavelli P, Shen Z (2016) Impedance-based analysis of grid-synchronization stability for three-phase paralleled converters. IEEE Trans Power Electron 31(1):675–687
5. Fang J, Lin P, Li H, Yang Y, Tang Y (2019) An improved virtual inertia control for three-phase voltage source converters connected to a weak grid. IEEE Trans Power Electron 34(9):8660–8670
6. Fang J, Zhang R, Li H, Tang Y (2019) Frequency derivative-based inertia enhancement by grid-connected power converters with a frequency-locked-loop. IEEE Trans Smart Grid 10(5):4918–4927
7. Kundur P (1994) Power system stability and control. McGraw-Hill, New York, NY, USA
8. Fang J, Xiao G, Yang X, Tang Y (2017) Parameter design of a novel series-parallel-resonant *LCL* filter for single-phase half-bridge active power filters. IEEE Trans Power Electron 32(1):200–217
9. Wang J, Yan JD, Jiang L, Zou J (2016) Delay-dependent stability of single-loop controlled grid-connected inverters with *LCL* filters. IEEE Trans Power Electron 31(1):743–757
10. Tang Y, Loh PC, Wang P, Choo FH, Gao F (2012) Exploring inherent damping characteristic of *LCL*-filters for three-phase grid-connected voltage source inverters. IEEE Trans Power Electron 27:1433–1443
11. Yang H, Fang J, Tang Y (2019) Exploration of time-delay effect on the stability of grid-connected power converters with virtual inertia. In: Proceedings of the ICPE 2019-ECCE Asia, Bexco, Busan, Korea, 27–30 May 2019
12. Liserre M, Teodorescu R, Blaabjerg F (2006) Stability of photovoltaic and wind turbine grid-connected inverters for a large set of grid impedance values. IEEE Trans Power Electron 21:263–272
13. Fang J, Li X, Yang X, Tang Y (2017) An integrated trap−*LCL* filter with reduced current harmonics for grid-connected converters under weak grid conditions. IEEE Trans Power Electron 32(11):8446–8457
14. Erikson RW, Ćuk S, Middlebrook RD (1982) Large-signal modeling and analysis of switching regulators. In: IEEE PESC, pp 240–250
15. Wang X, Harnefors L, Blaabjerg F (2018) Unified impedance model of grid-connected voltage-source converters. IEEE Trans Power Electron 33(2):1775–1787
16. Wang X, Blaabjerg F (2019) Harmonic stability in power electronic-based power systems: concept, modelling, and analysis. IEEE Trans Smart Grid 10(3):2858–2870

17. Fang J, Li H, Tang Y, Blaabjerg F (2019) On the inertia of future more-electronics power systems. IEEE J Emerg Sel Topics Power Electron 7(4):2130–2146
18. AEMO (2017) International review of frequency control adaptation. Aust. Energy Market Operator, Melbourne, VIC, Australia
19. Fang J, Li H, Tang Y, Blaabjerg F (2018) Distributed power system virtual inertia implemented by grid-connected power converters. IEEE Trans Power Electron 33(10):8488–8499
20. AEMO (2017) Black system South Australia 28 September 2016–final report. Aust. Energy Market Operator, Melbourne, VIC, Australia
21. Dörfler F, Bullo F (2011) On the critical coupling for Kuramoto oscillators. SIAM J Appl Dyn Syst 10:1070–1099
22. Dörfler F, Chertkov M, Bullo F (2013) Synchronization in complex oscillator networks and smart grids. Proc Natl Acad Sci 110(6):2005–2010

Chapter 5
Stability Improvement Techniques

5.1 Converter-Level Stability Improvement Techniques

In the previous chapter, we have introduced converter-level stability problems. Such stability problems are classified into improper controller designs, time delays, weak grids, and power quality problems. Correspondingly, this section provides converter-level stability improvement techniques, including proper controller design, time-delay reduction, weak grid techniques, and power quality conditioning. Notably, we have detailed power quality conditioning equipment in Chap. 3. Instead, we focus on the improvement of converter adaptivity to power quality problems in this section.

5.1.1 Proper Controller Design

First of all, we should guarantee the stable operations of four types of GTCs (modeled in Sect. 4.1) under normal operating conditions. To do so, we must design controllers (ac current, dc voltage, PLLs, ac voltage, and grid-supportive controllers) in a proper way. As mentioned, stability is ensured if and only if all closed-loop poles are in the left-half plane (or within the unity circle) for continuous (or discrete) systems. According to the classic control theory, we use stability margins (such as GMs and PMs) to evaluate the degree of stability for simple systems. In the design stage, we tend to leave sufficient stability margins and hence ensure stability. In complicated systems with a number of parameters, we can tune and stabilize systems through sensitivity analysis [1].

J. Fang, *More-Electronics Power Systems: Power Quality and Stability*, Power Systems, https://doi.org/10.1007/978-981-15-8590-6_5

5.1.1.1 Controller Tuning Through Sensitivity Analysis

To introduce *sensitivity analysis*, we build a state-space model as

$$\begin{aligned} \frac{dx}{dt} &= \mathbf{Ax} + \mathbf{Bu} \\ \mathbf{y} &= \mathbf{Cx} + \mathbf{Du} \end{aligned}. \tag{5.1}$$

Assuming that **A** has *n* different eigenvalues (denoted as λ_i), which satisfy

$$\mathbf{A}\varphi = \mathbf{A}\begin{bmatrix} \varphi_1 & \varphi_2 & \cdots & \varphi_n \end{bmatrix} = \begin{bmatrix} \lambda_1\varphi_1 & \lambda_2\varphi_2 & \cdots & \lambda_n\varphi_n \end{bmatrix} \text{ and} \tag{5.2}$$

$$\psi A = \begin{bmatrix} \psi_1 \\ \psi_2 \\ \vdots \\ \psi_n \end{bmatrix} \mathbf{A} = \begin{bmatrix} \lambda_1\psi_1 \\ \lambda_2\psi_2 \\ \vdots \\ \lambda_n\psi_n \end{bmatrix}, \tag{5.3}$$

where φ and ψ represent the matrices consisting of right and left eigenvectors, respectively.

Reorganizing (5.2) and (5.3), we derive that

$$\mathbf{A}\varphi = \begin{bmatrix} \varphi_1 & \varphi_2 & \cdots & \varphi_n \end{bmatrix} \begin{bmatrix} \lambda_1 & 0 & 0 & 0 \\ 0 & \lambda_2 & 0 & 0 \\ 0 & 0 & \ldots & 0 \\ 0 & 0 & 0 & \lambda_n \end{bmatrix} = \varphi A^* \text{ and} \tag{5.4}$$

$$\psi\mathbf{A} = \begin{bmatrix} \psi_1 \\ \psi_2 \\ \vdots \\ \psi_n \end{bmatrix} \mathbf{A} = \begin{bmatrix} \lambda_1 & 0 & 0 & 0 \\ 0 & \lambda_2 & 0 & 0 \\ 0 & 0 & \ldots & 0 \\ 0 & 0 & 0 & \lambda_n \end{bmatrix} \begin{bmatrix} \psi_1 \\ \psi_2 \\ \vdots \\ \psi_n \end{bmatrix} = \mathbf{A}^*\psi, \tag{5.5}$$

where $\mathbf{A}^*$ denotes a diagonal matrix. Combining (5.4) and (5.5), we have

$$\psi\mathbf{A}\psi^{-1} = \varphi^{-1}\mathbf{A}\varphi = \Lambda \Rightarrow \psi = \varphi^{-1}. \tag{5.6}$$

Through the following linear transformation, i.e.,

$$\mathbf{x} = \varphi\mathbf{z}, \tag{5.7}$$

we reorganize the state-space model in (1) as

$$\begin{aligned} \frac{dz}{dt} &= \psi A\varphi z + \psi Bu = \mathbf{A}^*\mathbf{z} + \mathbf{B}^*\mathbf{u} \\ \mathbf{y} &= \mathbf{C}\varphi\mathbf{z} + \mathbf{Du} = \mathbf{C}^*\mathbf{z} + \mathbf{D}^*\mathbf{u} \end{aligned}, \tag{5.8}$$

Next, we perform sensitivity analysis, which essentially evaluates the sensitivity of eigenvalues (or closed-loop poles) to system parameters. With a specific eigenvalue λ_i, we simplify (5.2) into

$$\mathbf{A}\boldsymbol{\varphi}_i = \lambda_i \boldsymbol{\varphi}_i. \tag{5.9}$$

Taking derivative on both sides with respect to the entry a_{kj}, we have

$$\frac{\partial \mathbf{A}}{\partial a_{kj}} \boldsymbol{\varphi}_i + \mathbf{A} \frac{\partial \boldsymbol{\varphi}_i}{\partial a_{kj}} = \frac{\partial \lambda_i}{\partial a_{kj}} \boldsymbol{\varphi}_i + \lambda_i \frac{\partial \boldsymbol{\varphi}_i}{\partial a_{kj}}. \tag{5.10}$$

Further, by multiplying $\boldsymbol{\psi}_i$ on both sides, (5.10) yields

$$\boldsymbol{\psi}_i \frac{\partial \mathbf{A}}{\partial a_{kj}} \boldsymbol{\varphi}_i + \boldsymbol{\psi}_i \mathbf{A} \frac{\partial \boldsymbol{\varphi}_i}{\partial a_{kj}} = \boldsymbol{\psi}_i \frac{\partial \lambda_i}{\partial a_{kj}} \boldsymbol{\varphi}_i + \boldsymbol{\psi}_i \lambda_i \frac{\partial \boldsymbol{\varphi}_i}{\partial a_{kj}} \Rightarrow \boldsymbol{\psi}_i \frac{\partial \mathbf{A}}{\partial a_{kj}} \boldsymbol{\varphi}_i = \boldsymbol{\psi}_i \frac{\partial \lambda_i}{\partial a_{kj}} \boldsymbol{\varphi}_i. \tag{5.11}$$

Substitution of $\boldsymbol{\psi}_i \boldsymbol{\varphi}_i = 1$ [see (5.6)] into the right part of (5.11), the later becomes

$$\frac{\partial \lambda_i}{\partial a_{kj}} = \boldsymbol{\psi}_i \frac{\partial \mathbf{A}}{\partial a_{kj}} \boldsymbol{\varphi}_i \Rightarrow \frac{\partial \lambda_i}{\partial a_{kj}} = \psi_{ik} \varphi_{ji}, \tag{5.12}$$

which mathematically describes the sensitivity of a specific eigenvalue to a certain coefficient. On the basis of (5.12), we can further derive the sensitivity of a specific eigenvalue to a system parameter p as

$$\frac{\partial \lambda_i}{\partial p} = \boldsymbol{\psi}_i \frac{\partial \mathbf{A}}{\partial p} \boldsymbol{\varphi}_i. \tag{5.13}$$

The *participation factor* lumps the sensitivities of eigenvalue(s) to all diagonal matrix entries into a vector or a matrix. Specifically, we express participation factors as

$$\mathbf{P} = \begin{bmatrix} \mathbf{p}_1 \ \mathbf{p}_2 \ \dots \ \mathbf{p}_n \end{bmatrix} = \begin{bmatrix} \frac{\partial \lambda_1}{\partial a_{11}} & \frac{\partial \lambda_2}{\partial a_{11}} & \cdots & \frac{\partial \lambda_n}{\partial a_{11}} \\ \frac{\partial \lambda_1}{\partial a_{22}} & \frac{\partial \lambda_2}{\partial a_{22}} & \cdots & \frac{\partial \lambda_n}{\partial a_{22}} \\ \vdots & \vdots & \ddots & \vdots \\ \frac{\partial \lambda_1}{\partial a_{nn}} & \frac{\partial \lambda_2}{\partial a_{nn}} & \cdots & \frac{\partial \lambda_n}{\partial a_{nn}} \end{bmatrix} = \begin{bmatrix} \psi_{11}\varphi_{11} & \psi_{21}\varphi_{12} & \cdots & \psi_{n1}\varphi_{1n} \\ \psi_{12}\varphi_{21} & \psi_{22}\varphi_{22} & \cdots & \psi_{n2}\varphi_{2n} \\ \vdots & \vdots & \ddots & \vdots \\ \psi_{1n}\varphi_{n1} & \psi_{2n}\varphi_{n2} & \cdots & \psi_{nn}\varphi_{nn} \end{bmatrix}. \tag{5.14}$$

Through sensitivity analysis, we can identify the decisive parameters of unstable poles and tune controllers (and/or system parameters) accordingly. For illustration, Fig. 5.1 shows the sensitivity analysis of a critical pole (the real part) of Type IV GTCs, where we find that K_{id} is one of the decisive parameters. By increasing K_{id}, we can shift the pole leftwards and enlarge stability margins. It should be kept in mind that the rated values of different parameters vary greatly. For clarity, per unit

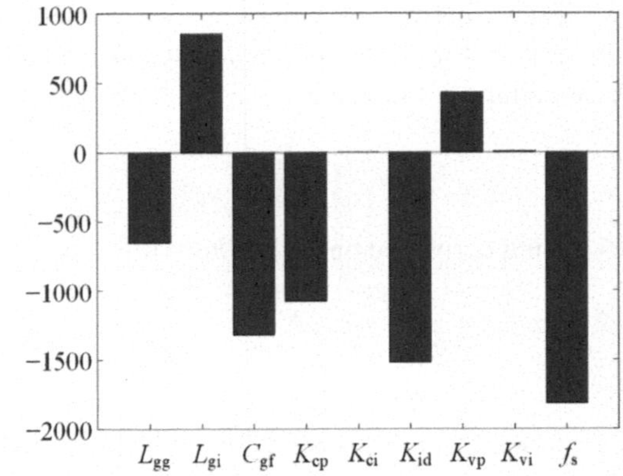

Fig. 5.1 Sensitivity analysis of a critical pole Type IV GTCs

sensitivity analysis is commonly used. In addition, we should remember that the shifting of one pole is intertwined with the others.

5.1.1.2 Passive Dampers

In parallel to controller tuning, we can facilitate controller designs through hardware-based techniques. One well-known technique is the employment of passive dampers.

Passive dampers consist of resistors and other passive components. They aim to attenuate resonances introduced by filters, GTCs, and/or SGs. Figure 5.2 demonstrates the schematic of series *RC* dampers, which are applied to *LCL* filters for resonance damping in this case [2]. The design of *RC* dampers is a trade-off between damping abilities and losses. In general, a larger resistor leads to better damping and more power losses. The opposite is true for a larger capacitor.

Figure 5.3 visualizes the damping effect of *RC* dampers through Bode diagrams of plant transfer functions [3]. Further validations of such passive dampers in various filters are validated in [3].

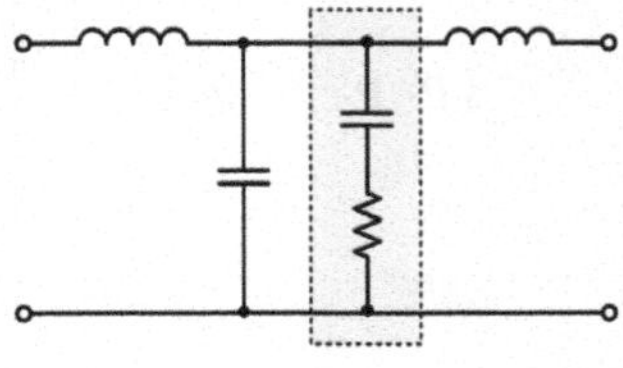

Fig. 5.2 Schematic of series *RC* passive dampers

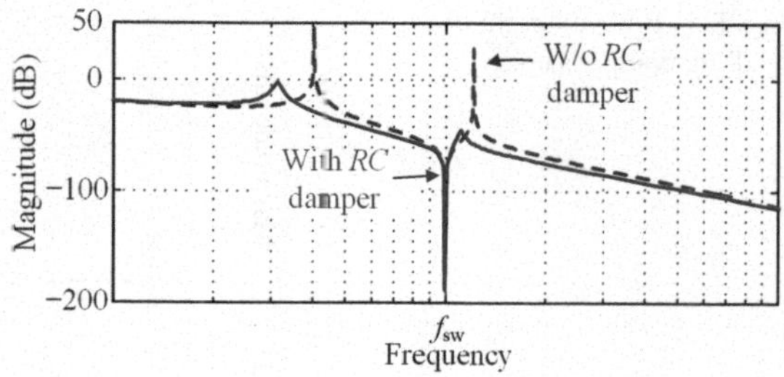

Fig. 5.3 Bode diagrams of plant functions with/without *RC* passive dampers [3]

5.1.2 Time-Delay Reduction

To reduce time delays, there are two methodologies. First, we can compensate time delays. The second option lies in the use of low- or no-latency controllers. This subsection will separately introduce them.

5.1.2.1 Time-Delay Compensation

We have discussed control and communication time delays in the previous chapter. As mentioned, both time delays are modeled as exponential functions e^{-sT_d}. To compensate time delays, a straightforward approach is to connect a compensator in series with the delay unit (such as the PWM generator). Ideally, the compensator should contain the inverse model of time delays, i.e.,

$$G_{com}(s) = e^{sT_d}. \tag{5.15}$$

To implement, we linearize the above model through Padé approximation as

$$G_{com}(s) = e^{sT_d} \approx \frac{1 + \frac{sT_d}{2}}{1 - \frac{sT_d}{2}}. \tag{5.16}$$

Clearly, the above compensator is unstable, as its pole is located in the right-half plane. Alternatively, we may linear the delay compensator as

$$G_{com}(s) = 1 + sT_d, \tag{5.17}$$

where the differential operator amplifies high-frequency noise, as the relevant magnitude is in proportion to the frequency.

To mitigate noise, we add a low-pass filter to the differential term, resulting in

$$G_{com}(s) = 1 + \frac{sT_d}{1 + sT_l} = \frac{1 + s(T_d + T_l)}{1 + sT_l}. \tag{5.18}$$

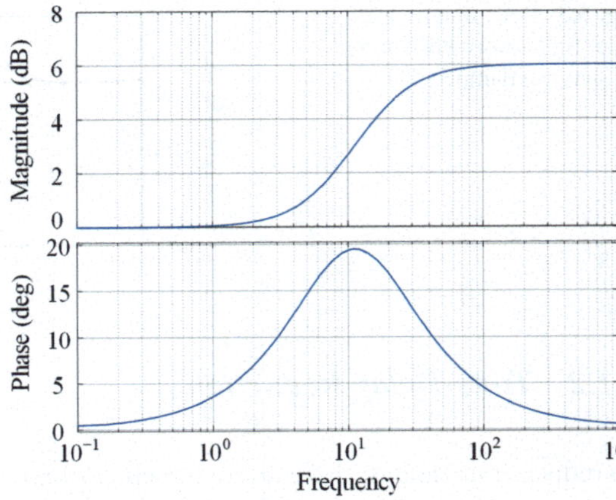

Fig. 5.4 Bode diagram of lead-lag compensators

In fact, the above model describes a lead-lag compensator. Figure 5.4 presents the Bode diagram of lead-lag compensators, where a positive phase angle is noticed around the center frequency. In essence, lead-lag compensators correct phase angles caused by time delays in a certain range, thereby stabilizing GTCs, as detailed in [4]. Notably, additional low-pass filters can be further incorporated to attenuate high-frequency noise, e.g.,

$$G_{\text{com}}(s) = \frac{1 + s(T_{\text{d}} + T_{\text{l}})}{1 + sT_{\text{l}}} \cdot \frac{1}{1 + sT_2}. \tag{5.19}$$

Note that the number of series stages determines roll-off rates in the high-frequency band.

As another example, we achieve differential operation at a certain frequency through high-pass filters, which are modeled as

$$G_{\text{com}}(s) = \frac{2\eta\omega_{\text{cr}}s^2}{s^2 + 2\eta\omega_{\text{cr}}s + \omega_{\text{cr}}^2}. \tag{5.20}$$

Around the angular frequency ω_{cr}, the second-order term s^2 and constant term ${\omega_{\text{cr}}}^2$ cancel out in the denominator. As a result, the compensator operates as a differentiator at ω_{cr}. Also, this differentiator is known as a second-order-generalized integrator (SOGI) [5]. Figure 5.5 shows the Bode diagrams of high-pass filters and pure differentiators. It is clear that the two diagrams intersect at ω_{cr}. Different from pure differentiators, high-pass filters do not amplify high-frequency noise.

5.1.2.2 Low-Latency Controllers

On top of time-delay compensation, we can reduce time delays from their sources. With this regard, low-latency controllers are preferable.

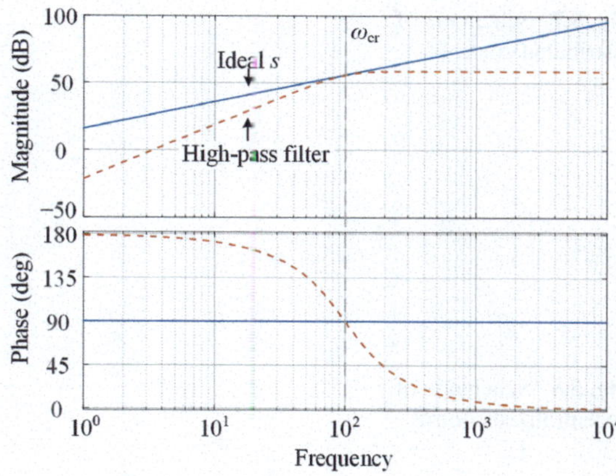

Fig. 5.5 Bode diagrams of high-pass filters and pure differentiators

As discussed in Sect. 1.2, FPGAs execute programs in a parallel fashion through configurable hardware. Therefore, FPGAs feature lower latency than sequentially executed controllers, such as DSPs. In time-demanding applications, we can employ FPGAs to reduce control time delays.

Regarding communication time delays, it is possible to reduce or even get rid of time delays through modified system communication and/or control architectures. For example, distributed control involves communications among multiple GTCs rather than GTCs and control centers [6]. In general, distributed control features reduced communication delays as compared with centralized control, particularly when only nearby units are communicated. It is even more desirable to remove communication time delays through decentralized control [7]. In this case, we control all the units based on their local information. As an example, droop control of individual SGs and GTCs belongs to decentralized control. However, systems without any communication may be subject to additional stability and reliability threatens. Figures 5.6, 5.7 and 5.8 visualize the schematics of centralized, distributed, and decentralized control, respectively. Notably, the circles and lines in Figs. 5.6, 5.7 and 5.8 denote GTCs and communication links, respectively. In more-electronics power systems, the penetration level of distributed sources keeps climbing up. To this end, distributed and decentralized control becomes increasingly popular. However, the advance of 5G communications will greatly reduce time delays and improve communication quality and hence add extra values to centralized control.

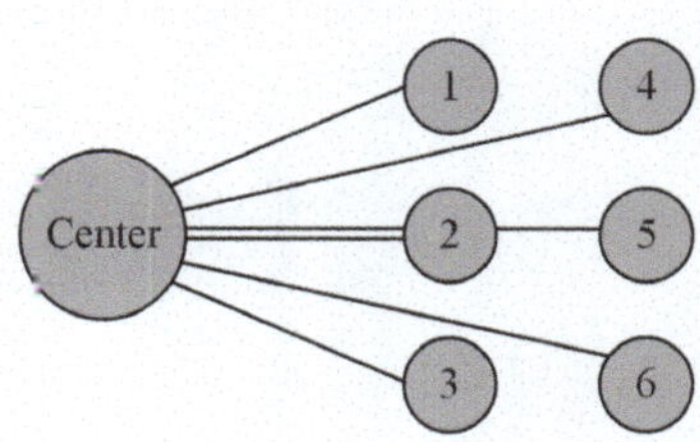

Fig. 5.6 Schematic of centralized control

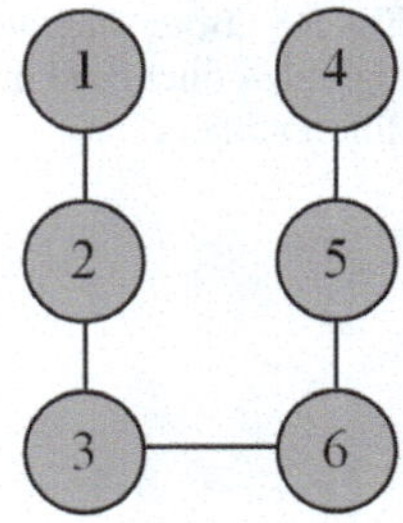

Fig. 5.7 Schematic of distributed control

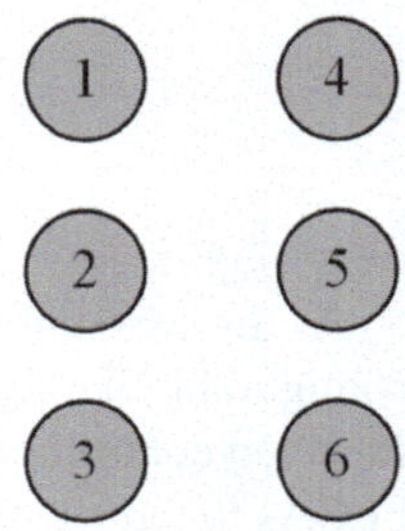

Fig. 5.8 Schematic of decentralized control

Finally, it should be noted that time delays may improve system stability, as mentioned in the previous chapter. In this scenario, we deliberately add extra time delays in the controller. For analog controllers, we can implement time delays according to Padé approximation. In terms of digital controllers, z^{-1} (or $z^{-\mathrm{N}}$) is a standard delay model. Taking Type III *LCL*-filtered GTCs with $f_{\mathrm{cr}} = f_{\mathrm{sw}}/6$ as an example, we note that the system is unstable without any additional technique. Figure 5.9 shows the relevant Bode diagrams with and without the additional delay. As observed, the additional delay helps to stabilize GTCs. Another example of using time delays is the repetitive controller, where we can find the model of time delays, i.e., $\mathrm{e}^{-sT_{\mathrm{o}}}$ [see (1.28)].

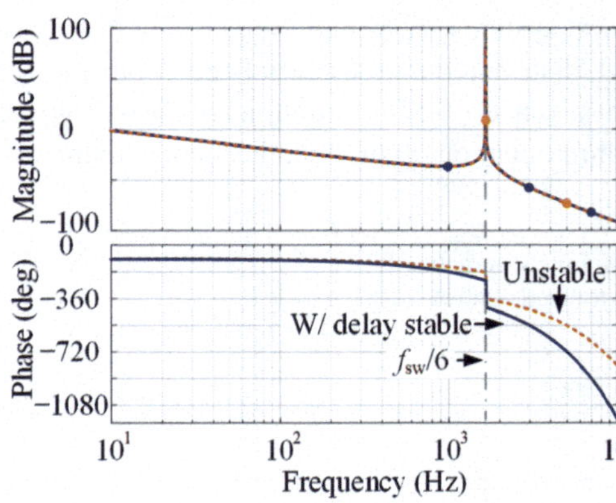

Fig. 5.9 Bode diagrams of Type III *LCL*-filtered GTCs with/without the additional delay

5.1.3 *Weak-Grid Techniques*

As stated in the previous chapter, weak-grid-induced stability problems are caused by parameter changes and local control loops. We will introduce their solutions in this subsection.

5.1.3.1 Robust Designs

Robust designs of systems and controllers are proven solutions to weak-grid-induced stability problems. Taking *LCL*-filtered ac grid-following GTCs (i.e., Types I and II GTCs) as an example, we express the relevant resonance frequency f_{cr} as

$$f_{cr} = \frac{1}{2\pi}\sqrt{\frac{L_{gi} + L_{gg} + L_s}{L_{gi}(L_{gg} + L_s)C_{gf}}}, \tag{5.21}$$

where L_s denotes the grid line inductance.

In weak grids, L_s can vary in a wide range. However, to ensure stability, f_{cr} must be maintained within certain boundaries. For grid-current-controlled (including single-loop or capacitor-current-damping) GTCs, the requirement of stable systems is ($f_s/2 > f_{cr} > f_s/6$) [8]. To comply with such a requirement, we derive the upper and lower limits of f_{cr} as

$$f_{cr\_max} = \frac{1}{2\pi}\sqrt{\frac{L_{gi} + L_{gg}}{L_{gi}L_{gg}C_{gf}}} \geq \frac{1}{2\pi}\sqrt{\frac{L_{gi} + L_{gg} + L_s}{L_{gi}(L_{gg} + L_s)C_{gf}}} \text{ and} \tag{5.22}$$

$$f_{cr\_min} = \frac{1}{2\pi}\sqrt{\frac{1}{L_{gi}C_{gf}}} < \frac{1}{2\pi}\sqrt{\frac{L_{gi} + L_{gg} + L_s}{L_{gi}(L_{gg} + L_s)C_{gf}}}, \tag{5.23}$$

respectively.

In essence, the upper limit (5.22) refers to the case of zero impedances, which is proved by

$$\frac{\partial f_{cr}}{\partial L_s} = \frac{1}{4\pi}\sqrt{\frac{(L_{gg} + L_s)L_{gi}C_{gf}}{L_{gg} + L_s + L_{gi}}} \cdot \frac{-L_{gi}^2 C_{gf}}{[(L_{gg} + L_s)L_{gi}C_{gf}]^2} < 0. \tag{5.24}$$

In addition, we prove the lower limit (5.23) through

$$f_{cr\_min} = \frac{1}{2\pi}\sqrt{\frac{L_{gi} + L_{gg} + L_s}{L_{gi}(L_{gg} + L_s)C_{gf}}} = \frac{1}{2\pi}\sqrt{\frac{1}{L_{gi}C_{gf}} + \frac{1}{(L_{gg} + L_s)C_{gf}}} > \frac{1}{2\pi}\sqrt{\frac{1}{L_{gi}C_{gf}}}, \tag{5.25}$$

which is essentially the resonance frequency of the case with infinitely large line impedances.

For robust designs, we select the upper and lower limits of resonance frequencies according to stability requirements. In the case of grid-current-controlled GTCs, these requirements become [8]

$$f_{cr\_max} = \frac{1}{2\pi}\sqrt{\frac{L_{gi}+L_{gg}}{L_{gi}L_{gg}C_{gf}}} < \frac{f_s}{2} \text{ and} \tag{5.26}$$

$$f_{cr\_min} = \frac{1}{2\pi}\sqrt{\frac{1}{L_{gi}C_{gf}}} > \frac{f_s}{6}. \tag{5.27}$$

The above conditions ensure system stability in weak grids.

Similar to robust system designs, robust controllers guarantee stable operation of GTCs with the consideration of system parameter changes. Noticeably, robust control has developed into a vivid branch of control engineering. As one example, robust control of GTCs in weak grids is detailed in [9]. However, to deal with parameter changes, robust controllers are often conservative.

5.1.3.2 Estimation and Adaptive Control

Another effective solution to weak-grid-induced stability problems is estimation and adaptive control. This technique works in two steps. First, we estimate grid information, such as line impedances, grid voltage amplitudes, and grid phase angles. Second, we adaptively change control schemes and/or parameters to get rid of instability.

Estimation Implementations

As mentioned, weak-grid-induced stability problems are mainly caused by either parameter changes or local control loops. Correspondingly, estimation objects are line impedances or grid voltage amplitudes and phase angles, respectively. Here, we achieve estimation by use of (active and reactive) power information [10].

Referring to the simplified schematic of two interconnected voltage sources in Fig. 4.27, we consider the first voltage source as a GTC (denoted by the subscript g) and the second one as the grid (denoted by the subscript s). Notably, their interconnected lines may contain not only grid line impedances but also part of converter impedances [10]. For three-phase systems, we rewrite the exchanged active and reactive power from (4.92) and (4.93) as

$$P_g = \frac{3}{2}\cdot\frac{R_s V_g(V_g - V_s\cos\delta_s) + X_s V_g V_s \sin\delta_s}{(R_s^2 + X_s^2)} \text{ and} \tag{5.28}$$

$$Q_g = \frac{3}{2}\cdot\frac{-R_s V_g V_s \sin\delta_s + X_s V_g(V_g - V_s\cos\delta_s)}{(R_s^2 + X_s^2)}, \tag{5.29}$$

where subscripts are updated with coefficients tripled. Clearly, there are line impedances and grid voltage information encoded in (5.28) and (5.29). As there are two equations, we can estimate two unknown variables.

We estimate the line inductance and resistance provided that the grid voltage amplitude V_s and phase angle δ_s are known. To remove power measurement noise, we adopt Kalman filters [11]. First, we build a discrete state-space model of line impedance estimation from (5.28) and (5.29) as

$$\begin{aligned} \mathbf{x_{k+1}} &= \mathbf{A_k x_k} \\ \mathbf{y_k} &= \mathbf{C_k x_k} + \mathbf{v_k} \end{aligned}, \tag{5.30}$$

where $\mathbf{v_k}$ denotes a random variable that models measurement noise. k represents the discrete time step. Note that $\mathbf{B_k}$ and $\mathbf{D_k}$ are removed from (5.30). $\mathbf{x_k}$ and $\mathbf{y_k}$ stand for the state and output vectors, which are expressed as

$$\mathbf{x_k} = \begin{bmatrix} x_{1k}(k) \\ x_{2k}(k) \end{bmatrix} = \begin{bmatrix} \frac{R_s(k)}{R_s(k)^2+X_s(k)^2} \\ \frac{X_s(k)}{R_s(k)^2+X_s(k)^2} \end{bmatrix}, \ \mathbf{y_k} = \begin{bmatrix} P_g(k) \\ Q_g(k) \end{bmatrix}. \tag{5.31}$$

In addition, $\mathbf{A_k}$ and $\mathbf{C_k}$ take the forms of

$$\mathbf{A_k} = \begin{bmatrix} 1 & 0 \\ 0 & 1 \end{bmatrix}, \ \mathbf{C_k} = \begin{bmatrix} C_{11}(k) & C_{12}(k) \\ -C_{12}(k) & C_{11}(k) \end{bmatrix}, \tag{5.32}$$

in which

$$C_{11}(k) = \frac{3V_g(k)\left[V_g(k) - V_s(k)\cos\delta_s(k)\right]}{2}, \ C_{12}(k) = \frac{3V_g(k)V_s(k)\sin\delta_s(k)}{2}. \tag{5.33}$$

Next, we introduce Kalman filtering schemes. Figure 5.10 depicts the flowchart of Kalman filters [10]. $\mathbf{x_{k|k-1}}$ and $\mathbf{x_k}^*$ represent the prediction and update variables of $\mathbf{x_k}$, respectively. $\mathbf{K_k}$ designates the Kalman gain. $\mathbf{P_{k|k-1}}$ and $\mathbf{P_k}$ refer to the estimate and updated error covariances, respectively. $\mathbf{Q_k}$ and $\mathbf{R_k}$ stand for the process and measurement noise covariance matrices, respectively. It is worth mentioning that $\mathbf{Q_k}$ and $\mathbf{R_k}$ are difficult to tune in complicated systems with numerous state variables. This is the reason why we propose a simplified second-order system modeled by (5.30). For simplicity, $\mathbf{Q_k}$ and $\mathbf{R_k}$ are designed to be constant matrices $\mathbf{Q_{(2\times 2)}}$ and $\mathbf{R_{(2\times 2)}}$. Notably, the Kalman filtering scheme works recursively in two steps, i.e., prediction and update [12]. In terms of initialization, there is no strict limitations as long as Kalman filters can converge.

According to the Type I GTC system and control parameters in Table 5.1, we initialize the vectors and matrices related to Kalman filters. For simplicity, the updated state $\mathbf{x_1}^*$ and updated error covariance $\mathbf{P_1}$ are randomly designed as unity vector and identity matrix, respectively. As the process covariance matrix $\mathbf{Q_{(2\times 2)}}$ plays a minor

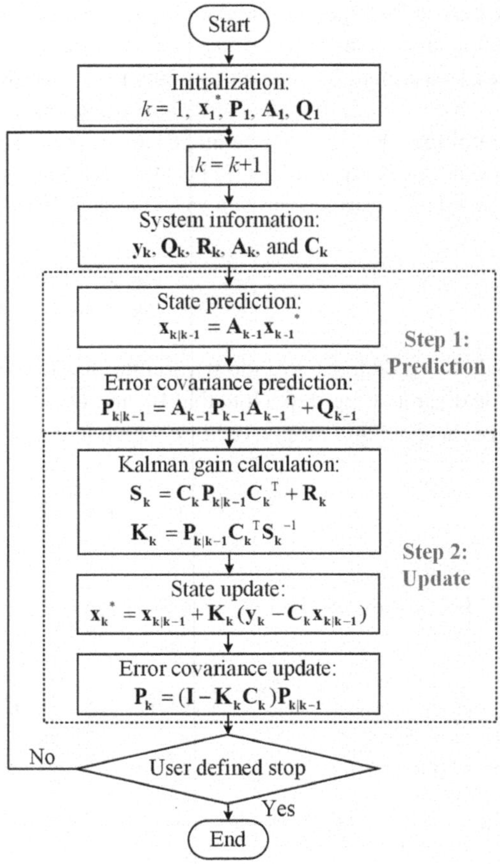

Fig. 5.10 Flowchart of Kalman filters [10]

role due to the exclusion of input matrix, we choose it as an identity matrix with a small gain factor, i.e., $\mathbf{Q}_{(2\times 2)} = [10^{-3}\ 0;0\ 10^{-3}]$. Importantly, the design of the measurement noise covariance matrix $\mathbf{R}_{(2\times 2)}$ is a tradeoff between dynamics and noise rejection. In this case, we set $\mathbf{R}_{(2\times 2)}$ as an identity matrix with a relatively large gain factor, i.e., $\mathbf{R}_{(2\times 2)} = [10^4\ 0;0\ 10^4]$.

Next, we derive the line inductance and resistance from the state variables as

$$R_s(k) = \frac{x_{1k}(k)}{x_{1k}(k)^2 + x_{2k}(k)^2},\ X_s(k) = \frac{x_{2k}(k)}{x_{1k}(k)^2 + x_{2k}(k)^2}. \tag{5.34}$$

On the contrary, we can estimate the grid voltage amplitude V_s and phase angle δ_s through (5.28) and (5.29) as long as we know the line impedance information. To do so, we rebuild the state space model as

$$\begin{aligned}\overline{\mathbf{x}}_{\mathbf{k}+1} &= \overline{\mathbf{A}}_{\mathbf{k}}\overline{\mathbf{x}}_{\mathbf{k}} \\ \overline{\mathbf{y}}_{\mathbf{k}} &= \overline{\mathbf{C}}_{\mathbf{k}}\overline{\mathbf{x}}_{\mathbf{k}} + \overline{\mathbf{v}}_{\mathbf{k}}\end{aligned}, \tag{5.35}$$

Table 5.1 Type I GTC system and control parameters

Descriptions	Symbols	Values
Converter filter inductance	L_{gg}	2 mH
Line inductance	L_s	10 mH
Line resistance	R_s	2 Ω
Dc-link voltage reference	V_{gdc_ref}	400 V
Grid voltage reference (rms)	V_{s_ref}	100 V
Current reference in the d-axis	I_{ggd_ref}	0 A
Current reference in the q-axis	I_{ggq_ref}	5 A
Fundamental frequency	f_o	50 Hz
Filter cut-off frequency	f_{cut}	50 Hz
Switching/sampling frequency	f_{sw}/f_s	10 kHz
Current control P gain	K_{cp}	50
Current control I gain	K_{ci}	500
PLL control P gain	K_{pllp}	0.5
PLL control I gain	K_{plli}	100

where the input and output vectors are changed into

$$\bar{\mathbf{x}}_{\mathbf{k}} = \begin{bmatrix} \bar{x}_{1k}(k) \\ \bar{x}_{2k}(k) \end{bmatrix} = \begin{bmatrix} V_s(k)\cos\delta_s(k) \\ V_s(k)\sin\delta_s(k) \end{bmatrix}, \quad \bar{\mathbf{y}}_{\mathbf{k}} = \begin{bmatrix} P_g(k) - \frac{3V_g(k)^2 R_s(k)}{2R_s(k)^2+2X_s(k)^2} \\ Q_g(k) - \frac{3V_g(k)^2 X_s(k)}{2R_s(k)^2+2X_s(k)^2} \end{bmatrix}. \tag{5.36}$$

The matrices associated with (5.35) are

$$\bar{\mathbf{A}}_{\mathbf{k}} = \begin{bmatrix} 1 & 0 \\ 0 & 1 \end{bmatrix}, \quad \bar{\mathbf{C}}_{\mathbf{k}} = \begin{bmatrix} \bar{C}_{11}(k) & \bar{C}_{12}(k) \\ -\bar{C}_{12}(k) & \bar{C}_{11}(k) \end{bmatrix}. \tag{5.37}$$

in which

$$\bar{C}_{11}(k) = \frac{-3V_g(k)R_s(k)}{2R_s(k)^2 + 2X_s(k)^2}, \quad \bar{C}_{12}(k) = \frac{3V_g(k)X_s(k)}{2R_s(k)^2 + 2X_s(k)^2}. \tag{5.38}$$

Once again, the estimation process follows Fig. 5 10. Finally, we obtain the grid voltage amplitude and phase angle as

$$V_s(k) = \sqrt{\bar{x}_{1k}(k)^2 + \bar{x}_{2k}(k)^2}, \; \delta_s(k) = \arctan\frac{\bar{x}_{2k}(k)}{\bar{x}_{1k}(k)}. \tag{5.39}$$

The estimated voltage amplitude and phase angle are real grid information, which can guide controller designs, e.g., voltage feedforward control [13]. In what follows, we introduce adaptive control based on the estimated grid information.

Fig. 5.11 Schematic of adaptive current controllers

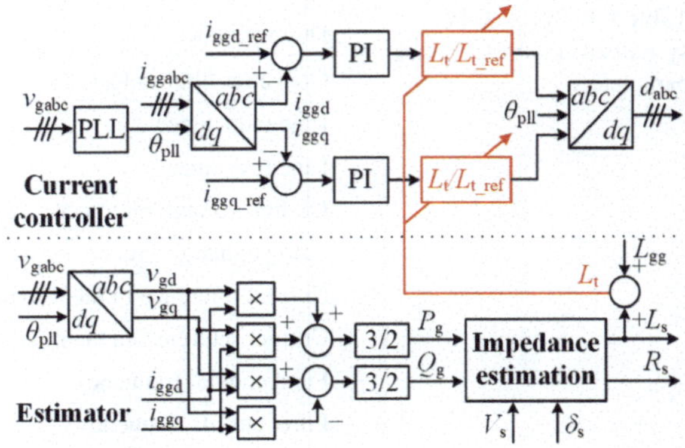

Adaptive Control

After having grid information, we can adaptively tune controllers to get rid of weak-grid-inducted stability problems. Let us take ac current controllers of Type I GTCs as an example. We adjust the current control gain in proportional to the line inductance to compensate the change of system plants.

Figure 5.11 shows the schematic of adaptive current controllers with line impedance estimation. Notably, the overall adaptive controller consists of an estimator and a current controller, where the estimator changes the current control gain for better stability. Active and reactive power measurement is referred to (2.25) and (2.26), respectively. The system and control parameters of Type I GTCs are listed in Table 5.1.

Figures 5.12 and 5.13 present the experimental waveforms of line impedance as well as grid voltage amplitude and phase angle estimation in the presence of considerable noise. Without Kalman filters, the estimated results are seriously distorted. Thanks to Kalman filters, we obtain accurate estimation results of grid information regardless of system noise.

Figures 5.14 and 5.15 provide the experimental waveforms of grid-injected currents and grid voltages under the same line impedance change from 10 to 2 mH for standard current control and adaptive current control, respectively. As analyzed

Fig. 5.12 Experimental waveforms of line impedance estimation

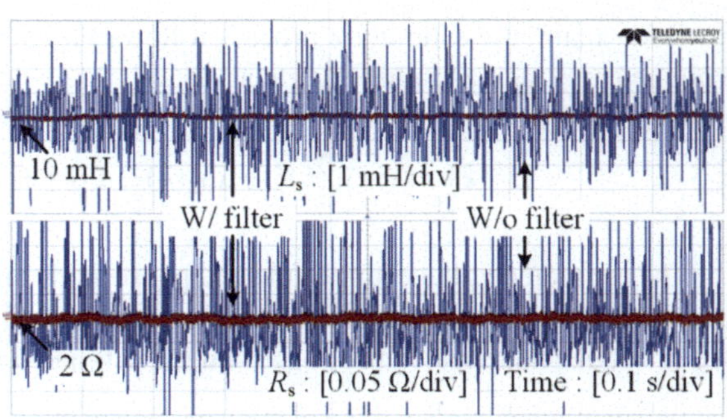

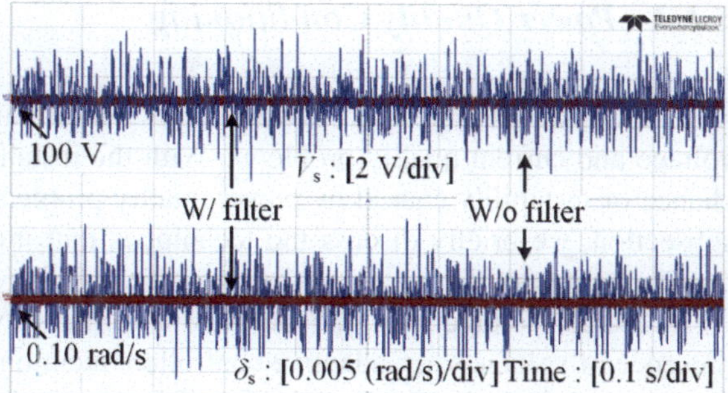

Fig. 5.13 Experimental waveforms of grid voltage amplitude and phase angle estimation

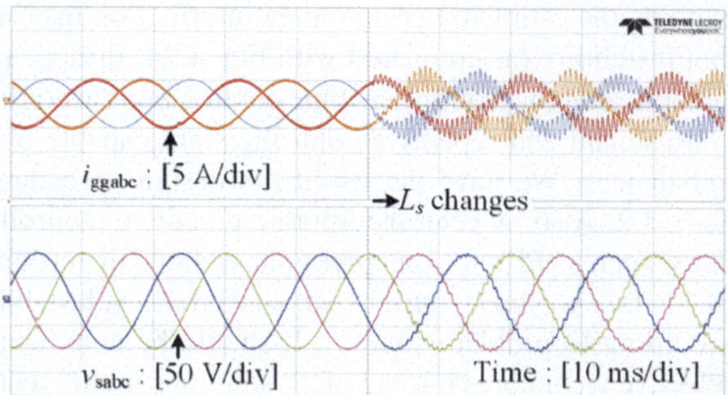

Fig. 5.14 Experimental waveforms of line impedance change with standard control

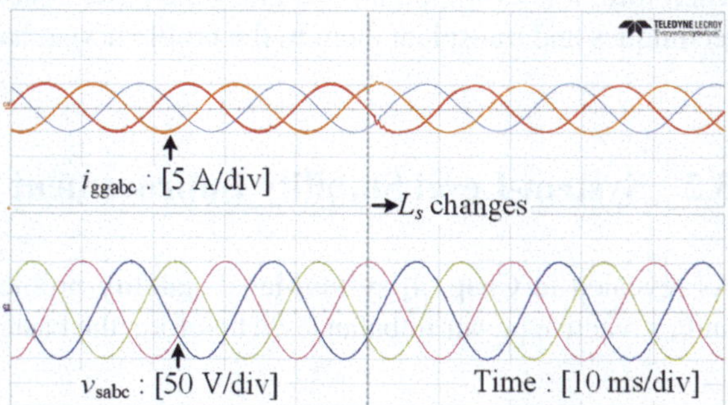

Fig. 5.15 Experimental waveforms of line impedance change with adaptive control

in the previous chapter, the reduced line inductance is equivalent to the amplification of current control gains, thereby leading to unstable oscillations in the case of standard current control. In contrast, the adaptive controller accurately estimates the line impedance and proportionally reduces its current gain, resulting in stable current control. Notably, instability disappears quickly, as the line impedance estimation settles down. In the experiment, the inductance step change was achieved by paralleling three additional inductors with the line inductors, which are emulated by discrete inductors. The concept of adaptive control can be transplanted into other controllers in the presence of weak grids.

5.1.4 *Power Quality Conditioning*

In Chap. 3, we have detailed power quality conditioning equipment that addresses voltage and current quality problems. With the installation of such equipment, the chance of instability caused by power quality problems will greatly reduce. In this subsection, we briefly discuss the techniques that improve GTC adaptability into power quality problems.

As stated, voltage sags are the most common power quality problems. In the presence of voltage sags, adaptive controllers mentioned in the previous subsection prove to be effective. To illustrate, Fig. 5.16 shows the experimental waveforms of voltage, current, and reactive power with the adaptive voltage droop control that exploits the estimated grid voltage information instead of PCC voltages. It is clear that instability (as compared with Fig. 4.24) disappears under a 5% voltage sag. In general, adaptive controllers tune control gains according to estimated grid voltages.

Resonant and repetitive controllers are capable of mitigating harmonics and/or imbalances. We have discussed the design procedure of repetitive controllers in Sect. 3.2. Also, we can incorporate repetitive controllers into grid synchronization units, such as PLLs, to suppress phase angle distortions.

So far, we have introduced converter-level stability improvement techniques according to problem sources. It should be commented that we have only shown selective techniques instead of all solutions in this section. Notably, numerous solutions have been developed, but most of which are application and problem specific. Therefore, it is of importance to grasp the basic concepts of stability improvement techniques and transplant them to the problem(s) at hand.

5.2 System-Level Stability Improvement Techniques

As revealed in Chap. 4, system-level stability problems are mainly due to active and/or reactive power imbalances. Therefore, the basic idea of system-level stability

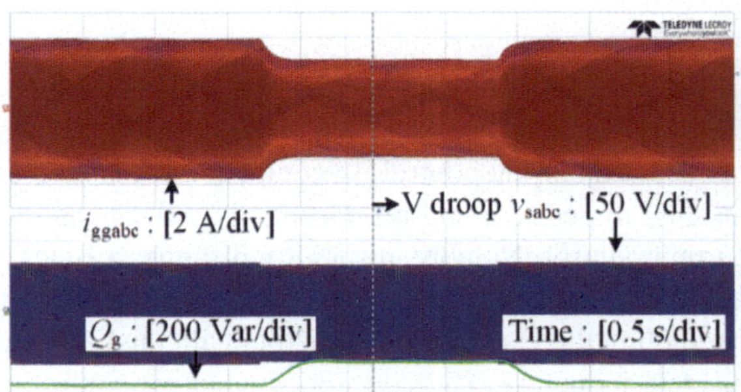

Fig. 5.16 Experimental waveforms of voltage droop with adaptive control

improvement lies in the balance of active and reactive power. In particular, we focus on the use of power converters for active and reactive power support.

5.2.1 *Active Power Balance*

It is claimed in the previous chapter that damping and/or inertia shortages are serious challenges of active power regulation. Accordingly, we tend to enhance damping and inertia in more-electronics power systems. As SGs provide both damping and inertia, we can achieve damping and inertia enhancement through the employment of redundant SGs. However, this solution is costly. In addition, it counteracts the general trend of the retirement of SGs. Therefore, we will exclude this solution in the following discussions.

5.2.1.1 Damping Enhancement Techniques

Referring back to Fig. 1.43, we notice the damping factor D in the transfer function from the power difference to the frequency, namely the per unit swing equation. In this part, we aim to increase the value of D.

To better illustrate the mechanism of damping enhancement, we redraw the swing equation block in Fig. 5.17, where the damping branch is relocated in the feedback branch. Clearly, the damping power Δp_{d_pu} is in proportion to the frequency Δf_{r_pu}. Therefore, to increase D, we make GTCs proportionally absorb active power in response to the detected frequency.

Next, we discuss control implementations of damping enhancement according to GTC types. Figure 5.18 shows the block diagram of grid-supportive control of Type

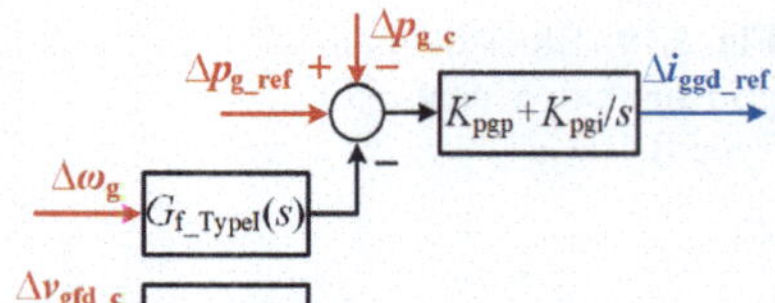

Fig. 5.17 Block diagram of swing equations with damping as feedback

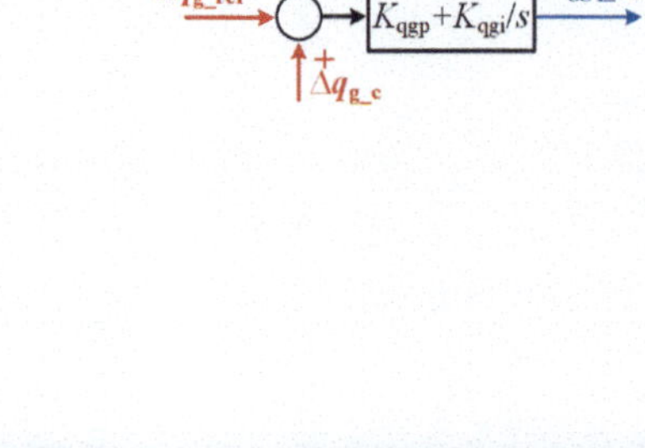

Fig. 5.18 Block diagram of grid-supportive control of Type I GTCs

I GTCs, where K_{pgp} and K_{pgi} (or K_{qgp} and K_{qgi}) denote the P and I gains of active power control (or reactive power control), respectively. $G_{f\_TypeI}(s)$ and $G_{v\_TypeI}(s)$ stand for the transfer functions of additional controllers for active and reactive power support, respectively. Notably, we express control parameters in terms of real values instead of per unit values. In this part, we focus on the damping enhancement that is related to active power control.

To achieve an increment of damping D_g, Type I GTCs detect the grid angular frequency $\Delta\omega_g$ and link it to their power references through $G_{f\_TypeI}(s)$. Due to the proportional relationship between the frequency and power, we design $G_{f\_TypeI}(s)$ as a P gain K_{f_damp}. It should be kept in mind that D_g and K_{f_damp} are referred to per unit and real values of damping, respectively. Considering the relevant rated quantities, D_g and K_{f_damp} have the following relationship

$$K_{f\_damp} = D_g \frac{VA_{g\_ref}}{2\pi f_0}, \tag{5.40}$$

where VA_{g_ref} represents the rated power of GTCs.

For Type II GTCs, Fig. 5.19 demonstrates the block diagram of grid-supportive control, where K_{dcp} and K_{dci} designate the P and I gains of dc voltage control, which replaces active power control. Similarly, $G_{f\_TypeII}(s)$ and $G_{v\_TypeII}(s)$ model the additional controllers for active and reactive power support, respectively. Recapping Fig. 4.5, we compare Δp_{g_c} and Δv_{g_dc} with the ignorance of low-pass filters. It is found that

$$\Delta p_{g\_c}(s) = \frac{\Delta v_{g\_dc}(s)}{C_{g\_dc} V_{g\_dc} s}. \tag{5.41}$$

Therefore, the proportional relationship between the frequency and active power translates into an integral relationship between the frequency and dc voltage. Correspondingly, $G_{f\_TypeII}(s) = K_{f\_damp}/(C_{g\_dc} V_{g\_dc} s)$ should be satisfied. Without loss of generality, we denote $G_{f\_TypeII}(s)$ as K_{f_damp}/s. It is worth mentioning that Type II GTCs feature dc capacitors instead of dc voltage sources. In consequence, they cannot continuously provide damping power (or frequency droop). Instead, Type II GTCs allow the improvement of damping around a frequency. In this case, a band pass filter should be inserted in series with K_{f_damp}/s, resulting in a low-pass filter.

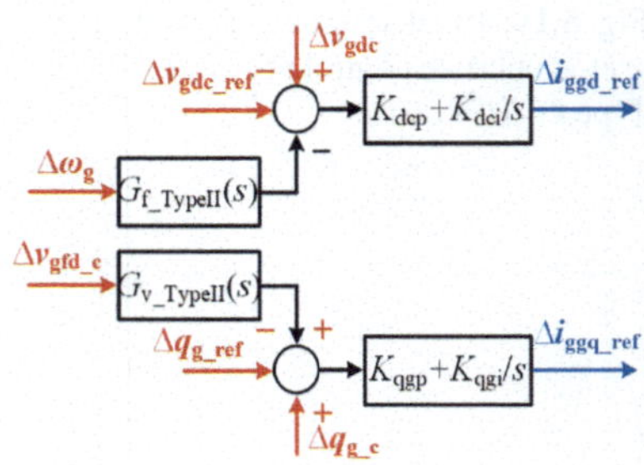

Fig. 5.19 Block diagram of grid-supportive control of Type II GTCs

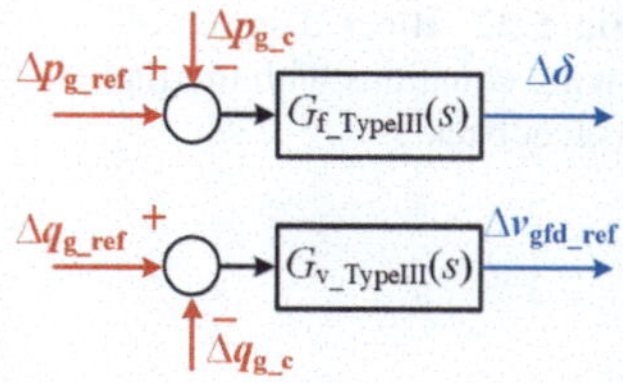

Fig. 5.20 Block diagram of grid-supportive control of Type III GTCs

Type III GTCs are ac grid-forming converters with dc voltage sources. Their power regulation framework is identical to that of SGs. For demonstration, Fig. 5.20 shows the block diagram of grid-supportive control of Type III GTCs, where $G_{f\_TypeIII}(s)$ and $G_{v\_TypeIII}(s)$ are related to the controllers for active and reactive power support, respectively. By comparing Figs. 1.43 and 5.20, we observe that the damping factor should appear in the denominator of $G_{f\_TypeIII}(s)$. In addition, we should include another integrator that changes the frequency into the phase angle. Given that $G_{f\_TypeIII}(s) = K_{f\_damp}/s$, we can derive the relationship between K_{f_damp} and the damping factor D_g as

$$K_{f\_damp} = \frac{2\pi f_0}{D_g V A_{g\_ref}}. \tag{5.42}$$

We should replace K_{f_damp} by a band-stop filter if damping is only exerted around a certain frequency.

Figure 5.21 presents the block diagram of grid-supportive control implemented by Type IV GTCs, where $G_{f\_TypeIV}(s)$ and $G_{v\_TypeIV}(s)$ refer to the controllers for active and reactive power support, respectively. Considering (5.41), we should design $G_{f\_TypeIV}(s)$ as $(C_{g\_dc} V_{g\_dc} K_{f\_damp})$ or K_{f_damp} to provide damping. Similar to Type II GTCs, Type IV GTCs feature dc capacitors and hence cannot continuously provide damping power. To perform damping around a certain frequency, $G_{f\_TypeIV}(s)$ should be designed as a band-stop filter.

We learn from the above discussions that damping (at 0 Hz) or frequency droop requires continuous active power injection or absorption. Therefore, the relevant requirement on the capacity of energy storage is high. As such, low energy–density units, such as capacitors, should be excluded from damping enhancement applications. Instead, batteries and flywheels are potential candidates. If damping is needed only around a certain ac frequency (e.g., oscillation damping), almost all energy

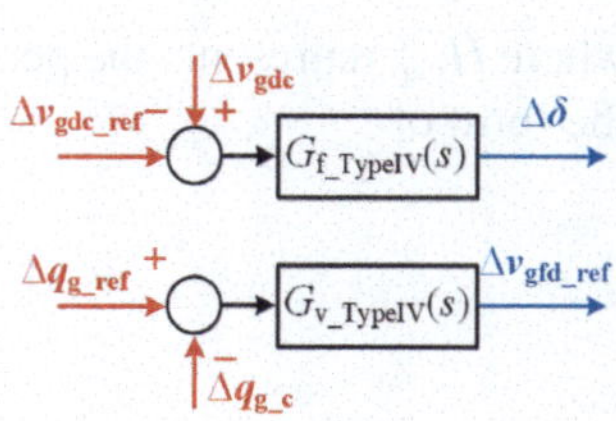

Fig. 5.21 Block diagram of grid-supportive control of Type IV GTCs

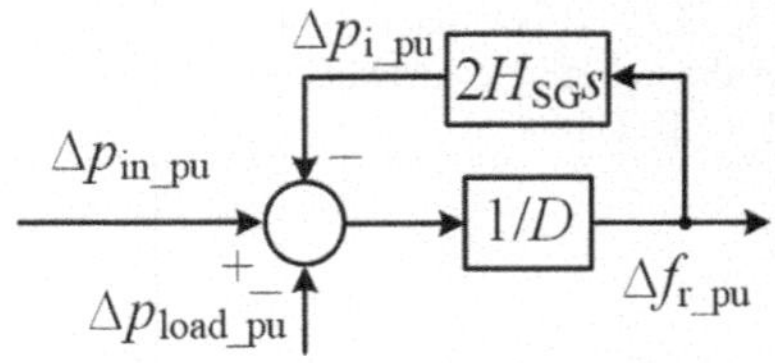

Fig. 5.22 Block diagram of swing equations with inertia as feedback

storage units can be applicable. Next, we proceed to show inertia enhancement techniques through GTCs.

5.2.1.2 Inertia Enhancement Techniques

Referring once again to Fig. 1.43, we reorganize the block diagram of the swing equation so that the inertia coefficient appears in the feedback branch, as shown in Fig. 5.22. In this case, the inertial power $\Delta p_{\mathrm{i\_pu}}$ is proportional to the derivative of the frequency $\Delta f_{\mathrm{r\_pu}}$. As such, we control the active power of GTCs according to the RoCoF (i.e., $\mathrm{d}f/\mathrm{d}t$) to generate virtual inertia $2H_{\mathrm{g}}$.

For Type I GTCs (see Fig. 5.18), we design $G_{\mathrm{f\_TypeI}}(s)$ as a differential controller, i.e., $G_{\mathrm{f\_TypeI}}(s) = sK_{\mathrm{f\_inertia}}$. Considering the relationship between real and per unit values, we derive

$$K_{\mathrm{f\_inertia}} = 2H_{\mathrm{g}} \frac{VA_{\mathrm{g\_ref}}}{2\pi f_0}. \tag{5.43}$$

The above equation provides a guidance for the design of virtual inertia controllers. However, pure differentiators are impractical due to the amplification of high frequency noise, as discussed. As a solution, frequency-locked loops (FLLs) can detect RoCoF signals in a fast and accurate way [5].

Regarding Type II GTCs, we refer to the control block diagram in Fig. 5.19. On the basis of (5.41), the virtual inertia controller takes the form of a proportional gain, i.e., $G_{\mathrm{f\_TypeII}}(s) = K_{\mathrm{f\_inertia}}/(C_{\mathrm{g\_dc}} V_{\mathrm{g\_dc}})$ or $K_{\mathrm{f\_inertia}}$. Further, we expand $G_{\mathrm{f\_TypeII}}(s)$ as

$$G_{\mathrm{f\_TypeII}}(s) = \frac{H_{\mathrm{g}} VA_{\mathrm{g\_ref}}}{\pi f_0 C_{\mathrm{g\_dc}} V_{\mathrm{g\_dc}}} = \frac{V_{\mathrm{g\_dc}} H_{\mathrm{g}}}{2\pi f_0 H_{\mathrm{cap}}}, \tag{5.44}$$

where H_{cap} represents the per unit electrical energy stored in a dc capacitor. It takes the form of

$$H_{\mathrm{cap}} = \frac{C_{\mathrm{g\_dc}} V_{\mathrm{g\_dc}}^2}{2VA_{\mathrm{g\_ref}}}. \tag{5.45}$$

We can design virtual inertia control according to (5.44) for Type II GTCs. However, it should be highlighted that inertia emulation causes dc voltage variations, which must be maintained within certain boundaries to avoid over modulation and voltage stresses. This poses an upper limit on the virtual inertia gain. The detailed tuning guidance of virtual inertia control can be found in [14]. Alternatively, we may design dc capacitances and voltages based on inertia synthesis requirements.

To deliver virtual inertia, we can derive the form of Type III GTC controllers by comparing Figs. 5.17 and 5.20. Specifically, $G_{\mathrm{f\_TypeIII}}(s) = K_{\mathrm{f\_inertia}}/s^2$ should be satisfied, where

$$K_{\mathrm{f\_inertia}} = \frac{2\pi f_0}{2H_{\mathrm{g}}VA_{\mathrm{g\_ref}}}. \tag{5.46}$$

Remember that one integrator in $G_{\mathrm{f\_TypeIII}}(s)$ is introduced by the frequency to phase angle transfer function, while the other one is shown in Fig. 5.17.

Referring to Fig. 5.21 and considering (5.41), we obtain the virtual inertia controller of Type IV GTCs as $G_{\mathrm{f\_TypeIV}}(s) = (C_{\mathrm{g\_dc}}V_{\mathrm{g\_dc}}K_{\mathrm{f\_inertia}})/s$ or $K_{\mathrm{f\_inertia}}/s$. Similar to Type II GTCs, Type IV GTCs must limit dc voltage variations according to dc capacitors and system requirements.

Table 5.2 summarizes the control implementations of grid-supportive services, where the gains are unified for simplicity. However, it should be noted that the same variable (e.g., $K_{\mathrm{f\_damp}}$) can have different implications for GTCs with different types.

Figure 5.23 demonstrates the experimental waveforms of frequency regulation with and without inertia enhancement by Type III GTCs. Obviously, the increment of

Table 5.2 Control implementations of grid-supportive services

Converters	Services	$G_f(s)$	$G_v(s)$
Type I	Dc damping	$K_{\mathrm{f\_damp}}$	$K_{\mathrm{v\_damp}}$
	Inertia	$sK_{\mathrm{f\_inertia}}$	$sK_{\mathrm{v\_inertia}}$
	Ac damping	Band-pass filter	Band-pass filter
Type II	Dc damping	$K_{\mathrm{f\_damp}}/s$	$K_{\mathrm{v\_damp}}$
	Inertia	$K_{\mathrm{f\_inertia}}$	$sK_{\mathrm{v\_inertia}}$
	Ac damping	Low-pass filter	Band-pass filter
Type III	Dc damping	$K_{\mathrm{f\_damp}}/s$	$K_{\mathrm{v\_damp}}$
	Inertia	$K_{\mathrm{f\_inertia}}/s^2$	$K_{\mathrm{v\_inertia}}/s$
	Ac damping	Band-stop filter/(s)	Band-stop filter
Type IV	Dc damping	$K_{\mathrm{f\_damp}}$	$K_{\mathrm{v\_damp}}$
	Inertia	$K_{\mathrm{f\_inertia}}/s$	$K_{\mathrm{v\_inertia}}/s$
	Ac damping	Band-stop filter	Band-stop filter

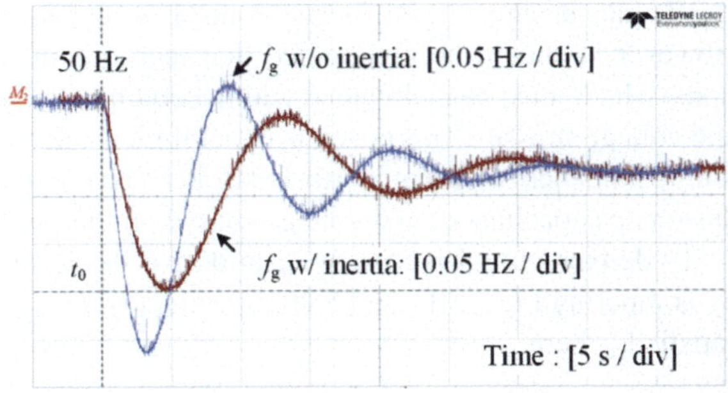

Fig. 5.23 Experimental waveforms of frequency regulation by Type III GTCs with and without inertia support

inertia improves frequency regulation in terms of frequency nadir and RoCoF. As for other types of GTCs, the relevant results of inertia enhancement can be referred to [4, 14–16]. Also, the concept of inertia emulation has been transplanted into multilevel converters and microgrids, as detailed in [17–19], respectively. The ideas of nonlinear inertia controllers are discussed in [20–22]. It is clear that inertia emulation through GTCs becomes popular in recent years.

In parallel to inertia from energy storage units in GTCs, other inertia enhancement techniques are under active research, as summarized in [16]. Among them, inertia from wind turbines is a well-demonstrated technique. So far, Ontario and Hydro-Quebec in Canada have mandated wind turbines for inertia services [23]. However, inertia from wind turbines is inferior to synchronous inertia. This inferiority is caused by a target conflict between inertia emulation and turbine speed regulation (or maximum power point tracking) [16]. After inertial response, wind turbines must absorb energy from the grid to achieve speed recovery, resulting in a secondary frequency dip or even pole slipping. Additionally, power reservation of primary sources illuminates some of the opportunities for inertia enhancement, as long as power curtailment has already been executed. Otherwise a non-trivial opportunity cost in the spilled energy deprives its economic advantage.

5.2.2 Reactive Power Balance

In Chap. 4, we have detailed reactive power compensation through power quality conditioning equipment, such as STATCOMs and UPQCs. In this subsection, we mainly focus on the reactive power supportive services that are dual to damping and inertia delivered by GTCs.

Referring back to Figs. 5.18, 5.19, 5.20 and 5.21, we observe that reactive power control (or voltage control) is dual to active power control (or frequency control). Therefore, we can design grid-supportive controllers related to reactive power in a similar way as before. However, it should be emphasized that the phase angle (rather than the frequency) corresponds to the voltage amplitude in terms of power exchanges. On the contrary, we normally treat the voltage amplitude as the frequency when designing damping or droop and inertia control. This explains the difference of controller forms in Type III GTCs (see Table 5.2). Notably, the reactive controllers of Types I and II GTCs (or Types III and IV GTCs) are identical. Another interesting thing of reactive power support is that GTCs with dc capacitors can absorb or inject reactive power continuously.

Before ending this chapter, we briefly look at stability problems due to grid-supportive services. Let us take Type II GTCs with virtual inertia control as an example to demonstrate how to conduct stability analysis. Referring to Fig. 5.19, we can model the virtual inertia controller $G_{\mathrm{f\_TypeII}}(s)$ as

$$\Delta \mathbf{y}_{\mathbf{inertia}} = \begin{bmatrix} 1 & K_{\mathbf{f}\_\mathrm{inertia}} \end{bmatrix} \Delta \mathbf{u}_{\mathbf{inertia}}, \tag{5.47}$$

where

$$\Delta \mathbf{u}_{\mathbf{inertia}} = \begin{bmatrix} \Delta v_{\mathrm{gdc0\_ref}} & \Delta \omega_{\mathrm{g}} \end{bmatrix}^{\mathrm{T}}, \Delta \mathbf{y}_{\mathbf{inertia}} = \Delta v_{\mathrm{gdc\_ref}}. \tag{5.48}$$

In (5.48), $\Delta v_{\mathrm{gdc0\_ref}}$ denotes the external dc voltage reference that is not affected by the inertia control. Recapping the model of Type II GTCs in Chap. 4 [see Fig. 4.10 and (4.55)], we further connect the virtual inertia controller to the GTC model, e.g., through the *lft* command in Matlab. After connection, we can continue to analyze stability as before. For demonstration, Fig. 5.24 illustrates the pole–zero maps of Type II GTCs with virtual inertia control, indicating unstable right-half-pane poles. Figure 5.25 demonstrates the relevant unstable experimental waveforms. We have discussed the corresponding solutions in Sect. 5.1.

In this chapter, we have introduced stability improvement techniques in more-electronics power systems. In the converter level, stability improvement techniques mainly include proper controller designs, time-delay reduction, weak grid information estimation and adaptive control, and power quality conditioning. In the system level, we improve active power balance through damping and inertia enhancement techniques. In addition, reactive power support is briefly discussed.

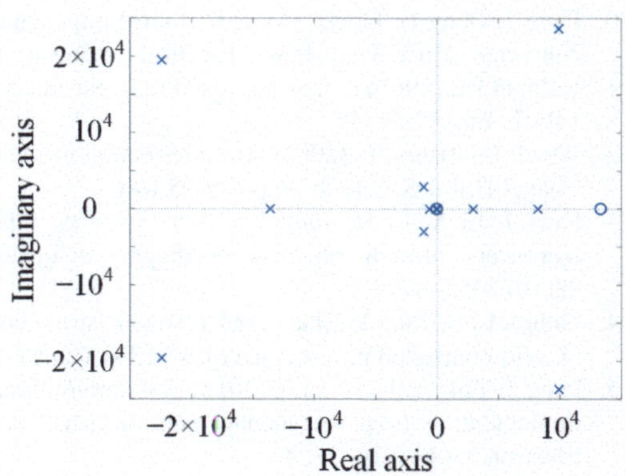

Fig. 5.24 Pole–zero maps of Type II GTCs with unstable virtual inertia control

Fig. 5.25 Experimental results of an unstable Type II GTC with virtual inertia control

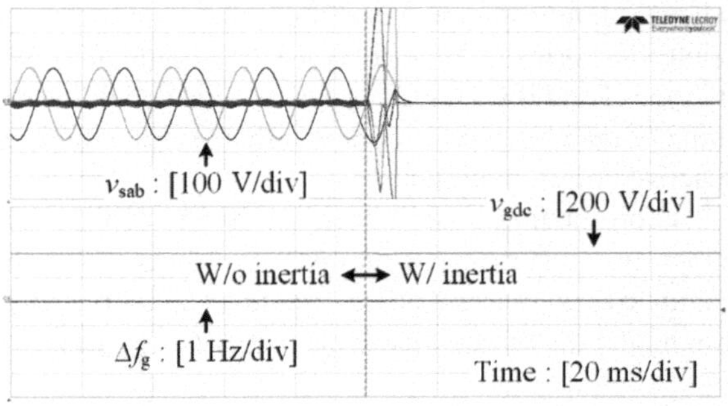

References

1. Kundur P (1994) Power system stability and control. McGraw-Hill, New York, NY, USA
2. Channegowda P, John V (2010) Filter optimization for grid interactive voltage source inverters. IEEE Trans Ind Electron 57(12):4106–4114
3. Fang J, Li X, Yang X, Tang Y (2017) An integrated trap−*LCL* filter with reduced current harmonics for grid-connected converters under weak grid conditions. IEEE Trans Power Electron 32(11):8446–8457
4. Fang J, Lin P, Li H, Yang Y, Tang Y (2019) An improved virtual inertia control for three-phase voltage source converters connected to a weak grid. IEEE Trans Power Electron 34(9):8660–8670
5. Fang J, Zhang R, Li H, Tang Y (2019) Frequency derivative-based inertia enhancement by grid-connected power converters with a frequency-locked-loop. IEEE Trans Smart Grid 10(5):4918–4927
6. Dorfler F, Simpson-Porco JW, Bullo F (2016) Breaking the hierarchy: distributed control and economic optimality in microgrids. IEEE Trans Control Netw Syst 3(3):241–253
7. Guerrero JM, Vasquez JC, Matas J, Vicuna LG, Castilla M (2011) Hierarchical control of droop-controlled AC and DC microgrids–a general approach toward standardization. IEEE Trans Ind Electron 58(1):158–172
8. Fang J, Xiao G, Yang X, Tang Y (2017) Parameter design of a novel series-parallel-resonant *LCL* filter for single-phase half-bridge active power filters. IEEE Trans Power Electron 32(1):200–217
9. Maccari LA, Massing JR, Schuch L, Rech C, Pinheiro H, Oliveira RCLF et al (2014) LMI-based control for grid-connected converters with *LCL* filters under uncertain parameters. IEEE Trans Power Electron 29:3776–3785
10. Fang J, Deng H, Goetz SM (2020) Grid impedance estimation through grid-forming power converters. IEEE Trans Power Electron (in press)
11. Kalman RE (1960) A new approach to linear filtering and prediction problems. Trans ASME J Basic Eng 82:34–45
12. Welch G, Bishop G (2004) An introduction to the Kalman filter. University North Carolina, Chapel Hill, USA, Tech. Rep. TR 95-041.
13. Fang J, Li X, Li H, Tang Y (2018) Stability improvement for three-phase grid-connected converters through impedance reshaping in quadrature-axis. IEEE Trans Power Electron 33(10):8365–8375
14. Fang J, Li H, Tang Y, Blaabjerg F (2018) Distributed power system virtual inertia implemented by grid-connected power converters. IEEE Trans Power Electron 33(10):8488–8499
15. Fang J, Tang Y, Li H, Li X (2018) A battery/ultracapacitor hybrid energy storage system for implementing the power management of virtual synchronous generators. IEEE Trans Power Electron 33(4):2820–2824

16. Fang J, Li H, Tang Y, Blaabjerg F (2019) On the inertia of future more-electronics power systems. IEEE J Emerg Sel Topics Power Electron 7(4):2130–2146
17. Yang S, Fang J, Tang Y, Qiu H, Dong C, Wang P (2019) Modular multilevel converter synthetic inertia-based frequency support for medium-voltage microgrids. IEEE Trans Ind Electron 66(11):8992–9002
18. Qiu H, Fang J, Tang Y (2018) Explore the capability of power electronic converters in providing power system virtual inertia. In: Proceedings of the IEEE ECCE, Portland, Oregon, USA, 23–27 Sept 2018
19. Zhang Z, Fang J, Tang Y (2019) A hybrid AC/DC microgrid with bidirectional virtual inertia support. In: Proceedings of the IEEE IFEEC, Singapore, 25–28 Nov 2019
20. Lu N, Fang J, Tang Y, Hredzak B (2019) A frequency deadband-based virtual inertia control for grid-connected power converters. In: Proceedings of the ICPE 2019-ECCE Asia, Bexco, Busan, Korea, 27–30 May 2019
21. Guo K, Fang J, Tang Y (2018) Autonomous DC-link voltage restoration for grid-connected power converters providing virtual inertia. In: Proceedings of the IEEE ECCE, Portland, Oregon, USA, 23–27 Sept 2018
22. Zhang R, Fang J, Tang Y (2019) Inertia emulation through supercapacitor energy storage systems. In: Proceedings of the ICPE 2019-ECCE Asia, Bexco, Busan, Korea, 27–30 May 2019
23. AEMO (2017) International review of frequency control adaptation. Aust. Energy Market Operator, Melbourne, VIC, Australia

The manufacturer's authorised representative in the EU is Springer Nature Customer Service Centre GmbH, Europaplatz 3, 69115 Heidelberg, Germany. If you have any concerns regarding our products, please contact ProductSafety@springernature.com

Printed and bound by CPI Group (UK) Ltd, Croydon, CR0 4YY
30/09/2026
02225559-0005